Advanced Information and Knowledge Processing

SpringerBriefs in Advanced Information and Knowledge Processing

Series editors

Xindong Wu, School of Computing and Informatics, University of Louisiana
at Lafayette, Lafayette, LA, USA
Lakhmi Jain, University of Canberra, Adelaide, SA, Australia

SpringerBriefs in Advanced Information and Knowledge Processing presents concise research in this exciting field. Designed to complement Springer's *Advanced Information and Knowledge Processing* series, this Briefs series provides researchers with a forum to publish their cutting-edge research which is not yet mature enough for a book in the *Advanced Information and Knowledge Processing* series, but which has grown beyond the level of a workshop paper or journal article. Typical topics may include, but are not restricted to:

Big Data analytics
Big Knowledge
Bioinformatics
Business intelligence
Computer security
Data mining and knowledge discovery
Information quality and privacy
Internet of things
Knowledge management
Knowledge-based software engineering
Machine intelligence
Ontology
Semantic Web
Smart environments
Soft computing
Social networks

SpringerBriefs are published as part of Springer's eBook collection, with millions of users worldwide and are available for individual print and electronic purchase. Briefs are characterized by fast, global electronic dissemination, standard publishing contracts, easy-to-use manuscript preparation and formatting guidelines and expedited production schedules to assist researchers in distributing their research fast and efficiently.

More information about this series at http://www.springer.com/series/16024

Rajendra Akerkar

Models of Computation for Big Data

Springer

Rajendra Akerkar
Western Norway Research Institute
Sogndal, Norway

ISSN 1610-3947   ISSN 2197-8441   (electronic)
Advanced Information and Knowledge Processing
ISSN 2524-5198   ISSN 2524-5201   (electronic)
SpringerBriefs in Advanced Information and Knowledge Processing
ISBN 978-3-319-91850-1   ISBN 978-3-319-91851-8   (eBook)
https://doi.org/10.1007/978-3-319-91851-8

Library of Congress Control Number: 2018951205

This Springer imprint is published by the registered company Springer Nature Switzerland AG
The registered company address is: Gewerbestrasse 11, 6330 Cham, Switzerland

Preface

This book addresses algorithmic problems in the age of big data. Rapidly increasing volumes of diverse data from distributed sources create challenges for extracting valuable knowledge and commercial value from data. This motivates increased interest in the design and analysis of algorithms for rigorous analysis of such data.

The book covers mathematically rigorous models, as well as some provable limitations of algorithms operating in those models. Most techniques discussed in the book mostly come from research in the last decade and of the algorithms we discuss have huge applications in Web data compression, approximate query processing in databases, network measurement signal processing and so on. We discuss lower bound methods in some models showing that many of the algorithms we presented are optimal or near optimal. The book itself will focus on the underlying techniques rather than the specific applications.

This book grew out of my lectures for the course on big data algorithms. Actually, *algorithmic aspects for modern data models* is a success in research, teaching and practice which has to be attributed to the efforts of the growing number of researchers in the field, to name a few Piotr Indyk, Jelani Nelson, S. Muthukrishnan, Rajiv Motwani. Their excellent work is the foundation of this book. This book is intended for both graduate students and advanced undergraduate students satisfying the discrete probability, basic algorithmics and linear algebra prerequisites.

I wish to express my heartfelt gratitude to my colleagues at Vestlandsforsking, Norway, and Technomathematics Research Foundation, India, for their encouragement in persuading me to consolidate my teaching materials into this book. I thank Minsung Hong for help in the LaTeX typing. I would also like to thank Helen Desmond and production team at Springer. Thanks to the INTPART programme funding for partially supporting this book project. The love, patience and encouragement of my father, son and wife made this project possible.

Sogndal, Norway   Rajendra Akerkar
May 2018

Contents

1 Streaming Models 1
   1.1   Introduction 1
   1.2   Space Lower Bounds 3
   1.3   Streaming Algorithms 4
   1.4   Non-adaptive Randomized Streaming 5
   1.5   Linear Sketch 5
   1.6   Alon–Matias–Szegedy Sketch 7
   1.7   Indyk's Algorithm 9
   1.8   Branching Program 11
         1.8.1   Light Indices and Bernstein's Inequality 14
   1.9   Heavy Hitters Problem 18
   1.10 Count-Min Sketch 19
         1.10.1 Count Sketch 21
         1.10.2 Count-Min Sketch and Heavy Hitters Problem 22
   1.11 Streaming k-Means 24
   1.12 Graph Sketching 25
         1.12.1 Graph Connectivity 27

2 Sub-linear Time Models 29
   2.1   Introduction 29
   2.2   Fano's Inequality 32
   2.3   Randomized Exact and Approximate Bound F_0 34
   2.4 t-Player Disjointness Problem 35
   2.5   Dimensionality Reduction 36
         2.5.1   Johnson Lindenstrauss Lemma 37
         2.5.2   Lower Bounds on Dimensionality Reduction 42
         2.5.3   Dimensionality Reduction for k-Means Clustering 45
   2.6   Gordon's Theorem 47
   2.7   Johnson–Lindenstrauss Transform 51
   2.8   Fast Johnson–Lindenstrauss Transform 55

2.9   Sublinear-Time Algorithms: An Example   58
2.10   Minimum Spanning Tree.   60
       2.10.1 Approximation Algorithm   62

3 Linear Algebraic Models   65
3.1   Introduction   65
3.2   Sampling and Subspace Embeddings   67
3.3   Non-commutative Khintchine Inequality.   70
3.4   Iterative Algorithms   71
3.5   Sarlós Method   72
3.6   Low-Rank Approximation   73
3.7   Compressed Sensing   77
3.8   The Matrix Completion Problem   79
       3.8.1 Alternating Minimization   81

4 Assorted Computational Models   85
4.1   Cell Probe Model   85
       4.1.1 The Dictionary Problem   86
       4.1.2 The Predecessor Problem   87
4.2   Online Bipartite Matching   89
       4.2.1 Basic Approach   89
       4.2.2 Ranking Method   90
4.3   MapReduce Programming Model.   91
4.4   Markov Chain Model   93
       4.4.1 Random Walks on Undirected Graphs   94
       4.4.2 Electric Networks and Random Walks.   95
       4.4.3 Example: The Lollipop Graph   95
4.5   Crowdsourcing Model.   96
       4.5.1 Formal Model   97
4.6   Communication Complexity   98
       4.6.1 Information Cost   98
       4.6.2 Separation of Information and Communication   99
4.7   Adaptive Sparse Recovery.   100

References   101

Chapter 1
Streaming Models

1.1 Introduction

In the analysis of big data there are queries that do not scale since they need massive computing resources and time to generate exact results. For example, count distinct, most frequent items, joins, matrix computations, and graph analysis. If approximate results are acceptable, there is a class of dedicated algorithms, known as streaming algorithms or sketches that can produce results orders-of magnitude faster and with precisely proven error bounds. For interactive queries there may not be supplementary practical options, and in the case of real-time analysis, sketches are the only recognized solution.

Streaming data is a sequence of digitally encoded signals used to represent information in transmission. For streaming data, the input data that are to be operated are not available all at once, but rather arrive as continuous data sequences. Naturally, a data stream is a sequence of data elements, which is extremely bigger than the amount of available memory. More often than not, an element will be simply an (integer) number from some range. However, it is often convenient to allow other data types, such as: multidimensional points, metric points, graph vertices and edges, etc. The goal is to approximately compute some function of the data using only one pass over the data stream. The critical aspect in designing data stream algorithms is that any data element that has not been stored is ultimately lost forever. Hence, it is vital that data elements are properly selected and preserved. Data streams arise in several real world applications. For example, a network router must process terabits of packet data, which cannot be all stored by the router. Whereas, there are many statistics and patterns of the network traffic that are useful to know in order to be able to detect unusual network behaviour. Data stream algorithms enable computing such statistics fast by using little memory. In Streaming we want to maintain a sketch $F(X)$ on the fly as X is updated. Thus in previous example, if numbers come on the fly, I can keep a running sum, which is a streaming algorithm. The streaming setting appears in a lot of places, for example, your router can monitor online traffic. You can sketch the number of traffic to find the traffic pattern.

© The Author(s), under exclusive license to Springer Nature Switzerland AG 2018
R. Akerkar, *Models of Computation for Big Data*, SpringerBriefs in Advanced
Information and Knowledge Processing, https://doi.org/10.1007/978-3-319-91851-8_1

The fundamental mathematical ideas to process streaming data are sampling and random projections. Many different sampling methods have been proposed, such as domain sampling, universe sampling, reservoir sampling, etc. There are two main difficulties with sampling for streaming data. First, sampling is not a powerful primitive for many problems since too many samples are needed for performing sophisticated analysis and a lower bound is given in. Second, as stream unfolds, if the samples maintained by the algorithm get deleted, one may be forced to resample from the past, which is in general, expensive or impossible in practice and in any case, not allowed in streaming data problems. Random projections rely on dimensionality reduction, using projection along random vectors. The random vectors are generated by space-efficient computation of random variables. These projections are called the sketches. There are many variations of random projections which are of simpler type.

Sampling and sketching are two basic techniques for designing streaming algorithms. The idea behind sampling is simple to understand. Every arriving item is preserved with a certain probability, and only a subset of the data is kept for further computation. Sampling is also easy to implement, and has many applications. Sketching is the other technique for designing streaming algorithms. Sketch techniques have undergone wide development within the past few years. They are particularly appropriate for the data streaming scenario, in which large quantities of data flow by and the the sketch summary must continually be updated rapidly and compactly. A sketch-based algorithm creates a compact synopsis of the data which has been observed, and the size of the synopsis is usually smaller than the full observed data. Each update observed in the stream potentially causes this synopsis to be updated, so that the synopsis can be used to approximate certain functions of the data seen so far. In order to build a sketch, we should either be able to perform a single linear scan of the input data (in no strict order), or to scan the entire stream which collectively build up the input. See that many sketches were originally designed for computations in situations where the input is never collected together in one place, but exists only implicitly as defined by the stream. Sketch $F(X)$ with respect to some function f is a *compression* of data X. It allows us computing $f(X)$ (with approximation) given access only to $F(X)$. A sketch of a large-scale data is a small data structure that lets you approximate particular characteristics of the original data. The exact nature of the sketch depends on what you are trying to approximate as well as the nature of the data.

The goal of the streaming algorithm is to make one pass over the data and to use limited memory to compute functions of x, such as the frequency moments, the number of distinct elements, the heavy hitters, and treating x as a matrix, various quantities in numerical linear algebra such as a low rank approximation. Since computing these quantities exactly or deterministically often requires a prohibitive amount of space, these algorithms are usually randomized and approximate.

Many algorithms that we will discuss in this book are randomized, since it is often necessary to achieve good space bounds. A *randomized algorithm* is an algorithm that can toss coins and take different actions depending on the outcome of those tosses. Randomized algorithms have several advantages over deterministic ones. Usually, randomized algorithms tend to be simpler than deterministic algorithms for the same task. The strategy of picking a random element to partition the problem into subproblems and recursing on one of the partitions is much simpler. Further, for some problems randomized algorithms have a better asymptotic running time than their deterministic one. Randomization can be beneficial when the algorithm faces lack of information and also very useful in the design of online algorithms that learn their input over time, or in the design of oblivious algorithms that output a single solution that is good for all inputs. Randomization, in the form of sampling, can assist us estimate the size of exponentially large spaces or sets.

1.2 Space Lower Bounds

Advent of cutting-edge communication and storage technology enable large amount of raw data to be produced daily, and subsequently, there is a rising demand to process this data efficiently. Since it is unrealistic for an algorithm to store even a small fraction of the data stream, its performance is typically measured by the amount of space it uses. In many scenarios, such as internet routing, once a stream element is examined it is lost forever unless explicitly saved by the processing algorithm. This, along with the complete size of the data, makes multiple passes over the data impracticable.

Let us consider the distinct elements problems to find the number of distinct elements in a stream, where queries and additions are allowed. We take s the space of the algorithm, n the size of the universe from which the elements arrive, and m the length of the stream.

Theorem 1.1 *There is no deterministic exact algorithm for computing number of distinct elements in $O(min n, m)$ space (Alon et al. 1999).*

Proof Using a streaming algorithm with space s for the problem we are going to show how to encode $\{0, 1\}^n$ using only s bits. Obviously, we are going to produce an injective mapping from $\{0, 1\}^n$ to $\{0, 1\}^s$. Hence, this implies that s must be at least n. We look for procedures such that $\forall x\, Dec(Enc(x)) = x$ and $Enc(x)$ is a function from $\{0, 1\}^n$ to $\{0, 1\}^s$.

In the encoding procedure, given a string x, devise a stream containing and add i at the end of the stream if $x_i = 1$. Then $Enc(x)$ is the memory content of the algorithm on that stream.

In the decoding procedure, let us consider each i and add it at the end of the stream and query then the number of distinct elements. If the number of distinct elements increases this implies that $x_i = 0$, otherwise it implies that $x_i = 1$. So we can recover x completely. Hence proved.

Now we show that approximate algorithms are inadequate for such problem.

Theorem 1.2 *Any deterministic F_0 algorithm that provides 1.1 approximation requires $\Omega(n)$ space.*

Proof Suppose we had a collection F fulfilling the following:

- $|F| \geq 2^{cn}$, for some constant $c < 1$.
- $\forall S \in F, |S| = \frac{n}{100}$
- $\forall S \neq T \in F, |S \cap T| \leq \frac{n}{2000} \leq \frac{1}{20}|S|$

Let us consider the algorithm to encode vectors $x_S \forall S \in F$, where x_S is the indicator vector of set S. The lower bound follows since we must have $s \geq cn$. The encoding procedure is similar as the previous proof.

In the decoding procedure, let us iterate over all sets and test for each set S if it corresponds to our initial encoded set. Further take at each time the memory contents of M of the streaming algorithm after having inserted initial string. Then for each S, we initialize the algorithm with memory contents M and then feed element i if $i \in S$. Suppose if S equals the initial encoded set, the number of distinct elements does increase slightly, whereas if it is not it almost doubles. Considering the approximation assurance of the algorithm we understand that if S is not our initial set then the number of distinct elements grows by $\frac{3}{2}$.

In order to confirm the existence of such a family of sets F, we partition n into $\frac{n}{100}$ intervals of length 100 each. To form a set S we select one number from each interval uniformly at random. Obviously, such a set has size exactly $\frac{n}{100}$. For two sets S, T selected uniformly at random as before let U_i be the random variable that equals 1 if they have the same number selected from interval i. So, $P[U_i = 1] = \frac{1}{100}$. Hence the anticipated size of the intersection is just $E \sum_{i=1}^{\frac{n}{100}} = \frac{n}{100} \cdot \frac{1}{100}$. The probability that this intersection is bigger than five times its mean is smaller than $e^{-c'n}$ for some constant c', by a standard Chernoff bound. Finally, by applying a union bound over all feasible intersections one can prove the result.

1.3 Streaming Algorithms

An important aspect of streaming algorithms is that these algorithms have to be approximate. There are a few things that one can compute exactly in a streaming manner, but there are lots of crucial things that one can't compute that way, so we have to approximate. Most significant aggregates can be approximated online. Many of these approximate aggregates can be computed online. There are two ways: (1) Hashing: which turns a pretty identity function into hash. (2) sketching: you can take a very large amount of data and build a very small sketch of the data. Carefully done, you can use the sketch to get values of interest. This in turn will find a good sketch. All of the algorithms discussed in this chapter use sketching of some kind and some use hashing as well. One popular streaming algorithm is HyperLogLog by Flajolet. Cardinality estimation is the task of determining the number of distinct elements in a data stream. While the cardinality can be easily computed using space linear in the cardinality, for several applications, this is totally unrealistic and requires too much memory. Therefore, many algorithms that approximate the cardinality while using less resources have been developed. HyperLogLog is one of them. These algorithms play an important role in network monitoring systems, data mining applications, as well as database systems. The basic idea is if we have n samples that are hashed and inserted into a $[0, 1)$ interval, those n samples are going to make $n + 1$ intervals. Therefore, the average size of the $n + 1$ intervals has to be $1/(n + 1)$. By symmetry, the average distance to the minimum of those hashed types is also going to be $1/(n + 1)$. Furthermore, duplicates values will go exactly on top of previous values, thus the n is the number of unique values we have inserted. For instance, if we have ten samples, the minimum is going to be right around $1/11$. HyperLogLog is shown to be near optimal among algorithms that are based on order statistics.

1.4 Non-adaptive Randomized Streaming

The non-trivial update time lower bounds for randomized streaming algorithms in the Turnstile Model was presented in (Larsen et al. 2014). Only a specific restricted class of randomized streaming algorithms, namely those that are non-adaptive could be bounded. Most well-known turnstile streaming algorithms in the literature are non-adaptive. Reference (Larsen et al. 2014) gives the non-trivial update time lower bounds for both randomized and deterministic turnstile streaming algorithms, which hold when the algorithms are non-adaptive.

Definition 1.1 A non-adaptive randomized streaming algorithm is an algorithm where it may toss random coins before processing any elements of the stream, and the words read from and written to memory are determined by the index of the updated element and the initially tossed coins, on any update operation.

These constraints suggest that memory must not be read or written to based on the current state of the memory, but only according to the coins and the index. Comparing the above definition to the sketches, a hash function chosen independently from any desired hash family can emulate these coins, enabling the update algorithm to find some specific words of memory to update using only the hash function and the index of the element to update. This makes the non-adaptive restriction fit exactly with all of the Turnstile Model algorithm. Both the Count-Min Sketch and the Count-Median Sketch are non-adaptive and support point queries.

1.5 Linear Sketch

Many data stream problems cannot be solved with just a sample. We can rather make use of data structures which, include a contribution from the entire input, instead of simply the items picked in the sample. For instance, consider trying to count the number of distinct objects in a stream. It is easy to see that unless almost all items are included in the sample, then we cannot tell whether they are the same or distinct. Since a streaming algorithm gets to see each item in turn, it can do better. We consider a *sketch* as compact data structure which summarizes the stream for certain types of query. It is a linear transformation of the stream: we can imagine the stream as defining a vector, and the algorithm computes the product of a matrix with this vector.

As we know a data stream is a sequence of data, where each item belongs to the universe. A data streaming algorithm takes a data stream as input and computes some function of the stream. Further, algorithm has access the input in a streaming fashion, i.e. algorithm cannot read the input in another order and for most cases the algorithm can only read the data once. Depending on how items in the universe are expressed in data stream, there are two typical models:

- *Cash Register Model*: Each item in stream is an item of universe. Different items come in an arbitrary order.
- *Turnstile Model*: In this model we have a multi-set. Every in-coming item is linked with one of two special symbols to indicate the dynamic changes of the data set. The turnstile model captures most practical situations that the dataset may change over time. The model is also known as dynamic streams.

We now discuss the turnstile model in streaming algorithms. In the turnstile model, the stream consists of a sequence of updates where each update either inserts an element or deletes one, but a deletion cannot delete an element that does not exist. When there are duplicates, this means that the multiplicity of any element cannot go negative.

In the model there is a vector $x \in \mathbb{R}^n$ that starts as the all zero vector and then a sequence of updates comes. Each update is of the form (i, Δ), where $\Delta \in \mathbb{R}$ and $i \in \{1, \ldots, n\}$. This matches to the operation $x_i \leftarrow x_i + \Delta$.

Given a function f, we want to approximate $f(x)$. For example, in the distinct elements problem Δ is always 1 and $f(x) = |i : x_i \neq 0$.

The well-known approach for designing turnstile algorithms is **linear sketching**. The idea is to preserve in memory $y = \Pi x$, where $\Pi \in \mathbb{R}^{m \times n}$, a matrix that is short and fat. We know that $m < n$, obviously much smaller. We can see that y is m-dimensional, so we can store it efficiently but if we need to store the whole Π in memory then we will not get space-wise better algorithm. Hence, there are two options in creating and storing Π.

- Π is deterministic and so we can easily compute Π_{ij} without keeping the whole matrix in memory.
- Π is defined by k-wise independent hash functions for some small k, so we can afford storing the hash functions and computing Π_{ij}.

Let Π^i be the ith column of the matrix Π. Then $\Pi_x = \sum_{i=1}^{n} \Pi^i x_i$. So by storing $y = \Pi x$ when the update (i, Δ) occures we have that the new y equals $\Pi(x + \Delta e_i) = \Pi x + \Delta \Pi^i$. The first summand is the old y and the second summand is simply multiple of the ith column of Π. This is how updates take place when we have a linear sketch.

Now let us consider Moment Estimation Problem (Alon et al. 1999). The problem of estimating (frequency) moments of a data stream has attracted a lot of attention since the inception of streaming algorithms. Suppose let $F_p = \|x\|_p^p = \sum_{i=1}^{p} |x_i|^p$. We want to estimate the space needed to solve the moment estimation problem as p changes. There is a transition point in complexity of F_p.

$0 \leq p \leq 2$, $poly(\frac{logn}{\varepsilon})$ space is achievable for $(1 + \varepsilon)$ approximation with $\frac{2}{3}$ success probability (Alon et al. 1999; Indyk 2006). For $p > 2$ then we need exactly $\Theta(n^{1-\frac{2}{p}} poly(\frac{logn}{\varepsilon}))$ bits of space for $(1 + \varepsilon)$ space with $\frac{2}{3}$ success probability (Bar-Yossef et al. 2004; Indyk and Woodruff 2005).

1.6 Alon–Matias–Szegedy Sketch

Streaming algorithms aim to summarize a large volume of data into a compact summary, by maintaining a data structure that can be incrementally modified as updates are observed. They allow the approximation of particular quantities. Alon–Matias–Szegedy (AMS) sketches (Alon et al. 1999) are randomized summaries of the data that can be used to compute aggregates such as the second frequency moment and sizes of joins. AMS sketches can be viewed as random projections of the data in the frequency domain on ± 1 pseudo-random vectors. The key property of AMS sketches is that the product of projections on the same random vector of frequencies of the join attribute of two relations is an unbiased estimate of the size of join of the relations. While a single AMS sketch is inaccurate, multiple such sketches can be computed and combined using averages and medians to obtain an estimate of any desired precision.

In particular, the AMS Sketch is focused on approximating the sum of squared entries of a vector defined by a stream of updates. This quantity is naturally related to the Euclidean norm of the vector, and so has many applications in high-dimensional geometry, and in data mining and machine learning settings that use vector representations of data. The data structure maintains a linear projection of the stream with a number of randomly chosen vectors. These random vectors are defined implicitly by simple hash functions, and so do not have to be stored explicitly. Varying the size of the sketch changes the accuracy guarantees on the resulting estimation. The fact that the summary is a linear projection means that it can be updated flexibly, and sketches can be combined by addition or subtraction, yielding sketches corresponding to the addition and subtraction of the underlying vectors.

A common feature of (Count-Min and AMS) sketch algorithms is that they rely on hash functions on item identifiers, which are relatively easy to implement and fast to compute.

Definition 1.2 H is a k-wise independent hash family if

$$\forall i_1 \neq i_2 \neq \cdots i_k \in [n] \; and \; \forall j_1, j_2, \ldots, j_k \in [m]$$

$$\Pr_{h \in H}[h(i_1) = j_1 \wedge \cdots \wedge h(i_k) = j_k] = \frac{1}{m^k}$$

There are two versions of the AMS algorithm. The faster version, based on the hashing is also referred to as fast AMS to distinguish it from the original "slower" sketch, since each update is very fast.

Algorithm:

1. Consider a random hash function $h : [n] \rightarrow \{-1, +1\}$ from a four-wise independent family.
2. Let $v_i = h(i)$.
3. Let $y = < v, x >$, output y^2.
4. y^2 is an unbiased estimator with variance big-Oh of the square of its expectation.
5. Sample y^2 $m_1 = O(\frac{1}{\varepsilon^2})$ independent times : $\{y_1^2, y_2^2, \ldots, y_{m_1}^2\}$. Use Chebyshev's inequality to obtain a $(1 \pm \varepsilon)$ approximation with $\frac{2}{3}$ probability.
6. Let $\overline{y} = \frac{1}{m_1} \sum_{i=1}^{m_1} y_i^2$.
7. Sample $\overline{y}$ $m_2 = O(\log(\frac{1}{\delta}))$ independent times : $\{\overline{y}_1, \overline{y}_2, \ldots, \overline{y}_{m_2}\}$. Take the median to get $(1 \pm \varepsilon)$-approximation with probability $1 - \delta$.

Each of the hash function takes $O(\log n)$ bits to store, and there are $O(\frac{1}{\varepsilon^2} \log(\frac{1}{\delta}))$ hash functions in total.

Lemma 1.1 $E[y^2] = \|x\|_2^2$.

Proof

$$E[y^2] = E[(< v, x >)^2]$$

$$= E\left[\sum_{i=1}^{n} v_i^2 x_i^2 + \sum_{i \neq j} v_i v_j x_i x_j\right]$$

$$= E\left[\sum_{i=1}^{n} v_i^2 x_i^2\right] + E\left[\sum_{i \neq j} v_i v_j x_i x_j\right]$$

$$= \sum_{i=1}^{n} x_i^2 + 0$$

$$= \|x\|_2^2$$

where $E[v_i v_j] = E[v_j] \cdot E[v_k] = 0$ since pair-wise independence.

Lemma 1.2 $E[(y^2 - E[y^2])^2] \leq 2\|x\|_2^4$.

Proof

$$E[(y^2 - E[y^2])^2] = E[(\sum_{i \neq j} v_i v_j x_i x_j)^2]$$

$$= E[4\sum_{i<j} v_i^2 v_j^2 x_i^2 x_j^2 + 4\sum_{i \neq j \neq k} v_i^2 v_j v_k x_i^2 x_j x_k + 24\sum_{i<j<k<l} v_i v_j v_k v_l x_i x_j x_k x_l]$$

$$= 4\sum_{i<j} x_i^2 x_j^2 + 4\sum_{i \neq j \neq k} E[v_i^2 v_j v_k x_i^2 x_j x_k] + 24E[\sum_{i<j<k<l} v_i v_j v_k v_l x_i x_j x_k x_l]$$

$$= 4\sum_{i<j} x_i^2 x_j^2 + 0 + 0$$

$$\leq 2\|x\|_2^4$$

where $E[v_i^2 v_j v_k] = E[v_j] \cdot [v_k] = 0$ since pair-wise independence,
and $E[v_i v_j v_k v_l] = E[v_i]E[v_j]E[v_k]E[v_l] = 0$ since four-wise independence.

In the next section we will present an idealized algorithm with infinite precision, given by Indyk (Indyk 2006). Though the sampling-based algorithms are simple, they cannot be employed for turnstile streams, and we need to develop other techniques.

Let us call a distribution $\mathbb{D}$ over $\mathbb{R}$ $p-stable$ if for $z_1, \ldots, z_n$ from this distribution and for all $x \in \mathbb{R}^n$ we have that $\sum_{i=1}^{n} z_i x_i$ is a random variable with distribution $\|x\|_p \mathbb{D}$. An example of such a distribution are the Gaussians for $p = 2$ and for $p = 1$ the Cauchy distribution, which has probability density function $pdf(x) = \frac{1}{\pi(x+1)^2}$. From probability theory, we know that the central limit theorem establishes that, in some situations, when independent random variables are added, their properly normalized sum tends toward a normal distribution even if the original variables themselves are not normally distributed. Hence, by the Central Limit Theorem an average of d samples from a distribution approaches a Gaussian as d goes to infinity.

1.7 Indyk's Algorithm

The Indyk's algorithm is one of the oldest algorithms which works on data streams. The main drawback of this algorithm is that it is a two pass algorithm, i.e., it requires two linear scans of the data which leads to high running time.

Let the ith row of Π be z_i, as before, where z_i comes from a p-stable distribution. Then consider $y_i = \sum_{j=1}^{n} z_{ij} x_j$. When a query arrives, output the median of all the y_i. Without loss of generality, let us suppose a p-stable distribution has median equal to 1, which in fact means that for z from this distribution $\mathbb{P}(-1 \leq z \leq 1) \leq \frac{1}{2}$.
Let $\Pi = \{\pi_{ij}\}$ be an $m \times n$ matrix where every element π_{ij} is sampled from a p-stable distribution, D_p. Given $x \in \mathbf{R}^n$, Indyk's algorithm (Indyk 2006) estimates the p-norm of x as

$$\|x\|_p \approx \text{median}_{i=1,\ldots,m} |y_i|,$$

where $y = \Pi x$.

In a turnstile streaming model, each element in the stream reflects an update to an entry in x. When an algorithm would maintain x in memory and calculates $\|x\|_p$ at the end, hence need $\Theta(n)$ space, Indyk's algorithm stores y and Π. Combined with a space-efficient way to produce Π we attain Superior space complexity.

Let us suppose Π is generated with D_p such that if $Q \sim D_p$ then $\text{median}(|Q|) = 1$. So, we assume the probability mass of D_p assigned to interval $[-1, 1]$ is $1/2$. Moreover, let $I_{[a,b]}(x)$ be an indicator function defined as

$$I_{[a,b]}(x) = \begin{cases} 1 & x \in [a, b], \\ 0 & \text{otherwise.} \end{cases}$$

Let Q_i be the ith row of Π. We have

$$y_i = \sum_{j=1}^{n} Q_{ij} x_j \sim ||x||_p D_p, \tag{1.1}$$

which follows from the definition of p-stable distributions and noting that Q_{ij}'s are sampled from D_p. This implies

$$\mathbf{E}\left[I_{[-1,1]}\left(\frac{y_i}{||x||_p} \right) \right] = \frac{1}{2}, \tag{1.2}$$

since $y_i/||x||_p \sim D_p$.

Moreover, it is possible to show that

$$\mathbf{E}\left[I_{[-1-\varepsilon,1+\varepsilon]}\left(\frac{y_i}{||x||_p} \right) \right] = \frac{1}{2} + \Theta(\varepsilon), \tag{1.3}$$

$$\mathbf{E}\left[I_{[-1+\varepsilon,1-\varepsilon]}\left(\frac{y_i}{||x||_p} \right) \right] = \frac{1}{2} - \Theta(\varepsilon). \tag{1.4}$$

Next, consider the following quantities:

$$F_1 = \frac{1}{m} \sum_{i}^{m} I_{[-1-\varepsilon,1+\varepsilon]}\left(\frac{y_i}{||x||_p} \right), \tag{1.5}$$

$$F_2 = \frac{1}{m} \sum_{i}^{m} I_{[-1+\varepsilon,1-\varepsilon]}\left(\frac{y_i}{||x||_p} \right). \tag{1.6}$$

F_1 represents the fraction of y_i's that satisfy $|y_i| \leq (1 + \varepsilon)||x||_p$, and likewise, F_2 represents the fraction of y_i's that satisfy $|y_i| \leq (1 - \varepsilon)||x||_p$. Using linearity of expectation property, we have $\mathbf{E}[F_1] = 1/2 + \Theta(\varepsilon)$ and $\mathbf{E}[F_2] = 1/2 - \Theta(\varepsilon)$. Therefore, the median of $|y_i|$ lies in

$$[(1 - \varepsilon)||x||_p, (1 + \varepsilon)||x||_p]$$

as desired.

Next step is to analyze the variance of F_1 and F_2. We have

$$\mathbf{Var}\left(F_1 \right) = \frac{1}{m^2} \times m \times (\text{variance of the indicator variable}). \tag{1.7}$$

Since variance of any indicator variable is not more than 1, $\mathbf{Var}(F_1) \leq \frac{1}{m}$. Likewise, $\mathbf{Var}(F_2) \leq \frac{1}{m}$. With an appropriate choice of m now we can trust that the median of $|y_i|$ is in the desired ε-range of $||x||_p$ with high probability.

Hence, Indyk's algorithm works, but independently producing and storing all mn elements of Π is computationally costly. To invoke the definition of p-stable distributions for Eq. 1.1, we need the entries in each row to be independent from one another. The rows need to be pairwise independent for calculation of variance to hold.

Let us assume $w_i = \sum_{j=1}^{n} W_{ij}x_j$ where W_{ij}'s are k-wise independent p-stable distribution samples.

$$\mathbf{E}\left[I_{[a,b]}\left(\frac{w_i}{||x||_p}\right)\right] \approx_\varepsilon \mathbf{E}\left[I_{[a,b]}\left(\frac{y_i}{||x||_p}\right)\right]. \tag{1.8}$$

If we can make this claim, then we can use k-wise independent samples in each row instead of fully independent samples to invoke the same arguments in the analysis above. This has been shown for $k = \Omega(1/\varepsilon^p)$ (Kane et al. 2010). With this technique, we can state Π using only $O(k \lg n)$ bits; across rows, we only need to use 2-wise independent hash function that maps a row index to a $O(k \lg n)$ bit seed for the k-wise independent hash function.

Indyk's approach for the L_p norm is based on the property of the median. However, it is possible to construct estimators based on other quantiles and they may even outperform the median estimator, in terms of estimation accuracy. However, since the improvement is marginal for our parameters settings, we stick to the median estimator.

1.8 Branching Program

A *branching programs* are built on directed acyclic graphs and work by starting at a source vertex and testing the values of the variables that each vertex is labeled with and following the appropriate edge till a sink is reached, and accepting or rejecting based on the identity of the sink. The program starts at an source vertex which is not part of the grid. At each step, the program reads S bits of input, reflecting the fact that space is bounded by S, and makes a decision about which vertex in the subsequent column of the grid to jump to. After R steps, the last vertex visited by the program represents the outcome. The entire input, which can be represented as a length-RS bit string, induces a distribution over the final states. Here we wish to generate the input string using fewer ($\ll RS$) random bits such that the original distribution over final states is well preserved. The following theorem addresses this idea.

Theorem 1.3 (Nisan 1992) *There exists $h : \{0, 1\}^t \to \{0, 1\}^{RS}$ for $t = O(S \lg R)$ such that*

$$\left|P_{x \sim U(\{0,1\}^{RS})}\{f(B(x)) = 1\} - P_{y \sim U(\{0,1\}^t)}\{f(B(h(y))) = 1\}\right| \leq \frac{1}{2^S}. \tag{1.9}$$

for any branching program B and any function $f : \{0, 1\}^S \to \{0, 1\}$.

The function h can simulate the input to the branching program with only t random bits such that it is almost impossible to discriminate the outcome of the simulated program from that of the original program.

A random sample x from $\{0, 1\}^S$ and add x at the root. Repeat the following procedure to create a complete binary tree. At each vertex, create two children and copy the string over to the left child. For the right child, use a random 2-wise independent hash function $h_j : [2^S] \to [2^S]$ chosen for the corresponding level of the tree and record the result of the hash. Once we reach R levels, output the concatenation of all leaves, which is a length-RS bit string. Since each hash function requires S random bits and there are $\lg R$ levels in the tree, this function uses $O(S \lg R)$ bits total.

One way to simulate randomized computations with deterministic ones is to build a pseudorandom generator, namely, an efficiently computable function g that can stretch a short uniformly random seed of s bits into n bits that cannot be distinguished from uniform ones by small space machines. Once we have such a generator, we can obtain a deterministic computation by carrying out the computation for every fixed setting of the seed. If the seed is short enough, and the generator is efficient enough, this simulation remains efficient. We will use Nisan's pseudorandom generator (PRG) to derandomize Π in Indyk's algorithm. Specifically, when the column indexed by x is required, Nisans generator takes x as the input and, together with the original, the generator outputs a sequence of pseudorandom sequences.

1. Initialize $c_1 \leftarrow 0, c_2 \leftarrow 0$
2. For $i = 1, \ldots, m$:

    a. Initialize $y \leftarrow 0$
    b. For $j = 1, \ldots, n$:
       i. Update $y \leftarrow y + \pi_{ij} x_j$
    c. If $y \le (1 + \varepsilon) \|x\|_p$, then increment c_1
    d. If $y \le (1 - \varepsilon) \|x\|_p$, then increment c_2

This procedure uses $O(\lg n)$ bits and is a branching algorithm that imitate the proof of correctness for Indyk's algorithm. The algorithm succeeded if and only if at the end of the computation $c_1 > \frac{m}{2}$ and $c_2 < \frac{m}{2}$. The only source of randomness in this program are the π_{ij}'s. We will apply Nisan's PRG to generate these random numbers. We invoke Theorem 1.3 with the algorithm given above as B and an indicator function checking whether the algorithm succeeded or not as f. See that the space bound is $S = O(\lg n)$ and the number of steps taken by the program is $R = O(mn)$, or $O(n^2)$ since $m \le n$. This means we can delude the proof of correctness of Indyk's algorithm by using $O(\lg^2 n)$ random bits to produce Π. Indyk's algorithm uses p-stable distributions which only exist for $p \in (0, 2]$. We shall consider a case when $p > 2$.

Theorem 1.4 $n^{1-2/p} poly(\frac{\lg n}{\varepsilon})$ *space is necessary and sufficient.*

Nearly optimal lower bound related details are discussed in (Bar-Yossef et al. 2004) and (Indyk and Woodruff 2005).

In this chapter we will discuss the algorithm of Andoni (Andoni 2012), which is based on (Andoni et al. 2011; Jowhari et al. 2011). We will focus on $\varepsilon = \Theta(1)$. In this algorithm, we let $\Pi = PD$. P is a $m \times n$ matrix, where each column has a single non-zero element that is either 1 or -1. D is a $n \times n$ diagonal matrix with $d_{ii} = u_i^{-1/p}$, where $u_i \sim \text{Exp}(1)$.

That is to say,

$$P\{u_i > t\} = \begin{cases} 1 & t \le 0, \\ e^{-t} & t > 0. \end{cases}$$

So, same as the $0 < p \le 2$ case, we will keep $y = \Pi x$, but we estimate $\|x\|_p$ with

$$\|x\|_p \approx \|y\|_\infty = \max_i |y_i|. \tag{1.10}$$

Theorem 1.5 $P\left\{\frac{1}{4}\|x\|_p \le \|y\|_\infty \le 4\|x\|_p\right\} \ge \frac{11}{20}$ *for* $m = \Theta(n^{1-2/p} \lg n)$.

Let $z = Dx$, which means $y = Pz$. To prove Theorem 1.5, we will begin by showing that $\|z\|_\infty$ delivers a good estimate and then prove that applying P to z maintains it.

Claim $P\left\{\frac{1}{2}\|x\|_p \le \|z\|_\infty \le 2\|x\|_p\right\} \ge \frac{3}{4}$.

Proof Let $q = \min\left\{\frac{u_1}{|x_1|^p}, \dots, \frac{u_n}{|x_n|^p}\right\}$. We have

$$P\{q > t\} = P\left\{\forall i, u_i > t|x_i|^p\right\} \tag{1.11}$$

$$= \prod_{i=1}^n e^{-t|x_i|^p} \tag{1.12}$$

$$= e^{-t\|x\|_p^p}, \tag{1.13}$$

which implies $q \sim \frac{\text{Exp}(1)}{\|x\|_p^p}$. Thus,

$$P\left\{\frac{1}{2}\|x\|_p \le \|z\|_\infty \le 2\|x\|_p\right\} = P\left\{\frac{1}{2^p}\|x\|_p^{-p} \le q \le 2^p\|x\|_p^{-p}\right\} \tag{1.14}$$

$$= e^{-\frac{1}{2^p}} - e^{-2^p} \tag{1.15}$$

$$\ge \frac{3}{4}, \tag{1.16}$$

for $p > 2$.

The following claim establishes that if we could maintain Q instead of y then we would have a better solution to our problem. However we can not store Q in memory because it's n-dimensional and $n \gg m$. Thus we need to analyze $PQ \in \mathbb{R}^m$.

Claim *Let $Q = DX$. Then*

$$P\left(\frac{1}{2}\|x\|_p \leq \|Q\|_\infty \leq 2\|x\|_p\right) \geq 3/4$$

Let us suppose each entry in y is a sort of counter and the matrix P takes each entry in Q, hashes it to a random counter, and adds that entry of Q times a random sign to the counter. There will be collision because $n > m$ and only m counters. These will cause different Q_i to potentially cancel each other out or add together in a way that one might expect to cause problems. We shall show that there are very few large Q_i's.

Interestingly, small Q_i's and big Q_i's might collide with each other. When we add the small Q_i's, we multiply them with a random sign. So the expectation of the aggregate contributions of the small Q_i's to each bucket is 0. We shall bound their variance as well, which will show that if they collide with big Q_i's then with high probability this would not considerably change the admissible counter. Ultimately, the maximal counter value (i.e., $\|y\|_\infty$) is close to the maximal Q_i and so to $\|x\|_p$ with high probability.

1.8.1 Light Indices and Bernstein's Inequality

Bernstein's inequality in probability theory is a more precise formulation of the classical Chebyshev inequality in probability theory, proposed by S.N. Bernshtein in 1911; it permits one to estimate the probability of large deviations by a monotone decreasing exponential function. In order to analyse the light indices, we will use *Bernstein's inequality*.

Theorem 1.6 (Bernstein's inequality) *Suppose $R_1, \ldots, R_n$ are independent, and for all i, $|R_i| \leq K$, and $var(\sum_i R_i) = \sigma^2$. Then for all $t > 0$*

$$P\left(\left|\sum_i R_i - E\left(\sum_i R_i\right)\right| > t\right) \lesssim e^{-ct^2/\sigma^2} + e^{-ct/K}$$

We consider that the light indices together will not distort the heavy indices. Let us parametrize P as follows and choose a function $h : [n] \to [m]$ as well as a function $\sigma : [n] \to \{-1, 1\}$. Then,

$$P_{ij} = \begin{cases} \sigma(j) & \text{if } h(j) = i \\ 0 & \text{else.} \end{cases}$$

Therefore, h states element of the column to make non-zero, and σ states which sign to use for column j.

The following light indices claim holds with constant probability that for all $j \in [m]$,

Claim

$$\left| \sum_{j \in E: h(j)=i} \sigma(j)Q_j \right| < T/10.$$

If y_i has no heavy indices then the magnitude of y_i is much less than T. Obviously, it would not hinder with estimate. If y_i assigned the maximal Q_j, then by previous claim that is the only heavy index assigned to y_i. Therefore, all the light indices assigned to y_i would not change it by more than $T/10$, and since Q_j is within a factor of 2 of T, y_i will still be within a constant multiplicative factor of T. If y_i assigned some other heavy index, then the corresponding Q_j is less than $2T$ since y_i is less than the maximal Q_j. This claim concludes that y_i will be at most $2.1T$.

Ultimately:

$$y_i = \sum_{j: h(j)=i} \sigma(j)Q_j$$

$$= \sum_{j \in E: h(j)=i} \sigma(j)Q_j + \sigma(j_{heavy})Q_{j_{heavy}}$$

where the second term is added only if y_i has heavy index. By the triangle inequality,

$$|y_i| \in Q_{j_{heavy}} \pm \left| \sum_{j \in E: h(j)=i} \sigma(j)Q_j \right|$$

$$= Q_{j_{heavy}} \pm T/10$$

Applying this to the bucket containing the maximal z_i shows that bucket of y should hold at least $0.4T$. Furthermore, by similar argument all other buckets should hold at most $2.1T$.

Proof Fix $i \in [m]$. Then for $j \in L$, define

$$\delta_j = \begin{cases} 1 & \text{if } h(j) = i \\ 0 & \text{else.} \end{cases}$$

Then

$$\sum_{j \in L} \delta_j \sigma(j)Q_j$$

We will call the jth term of the summand R_j and then use Bernstein's inequality.

1. We have $E(\sum R_j) = 0$, since the $\sigma(j)$ represent random signs.

2. We also have $K = T/(v \lg(n))$ since $|\delta_j| \leq 1$, $|\sigma(j)| \leq 1$, and we iterate over light indices so $|Q_j| \leq T/(v \lg(n))$.

It remains only to compute $\sigma^2 = \varepsilon(\sum_j R_j)$. If we condition on Q, then it implies that

$$\left(\sum_j R_j | Q \right) \leq \frac{\|Q\|_2^2}{m}$$

We need to consider the randomness of Q into account. We will merely prove that σ^2 is small with high probability over the choice of Q. We will do this by computing the unconditional expectation of σ^2 and then using Markov. Now

$$E\left(\|Q\|_2^2 \right) = \sum_j x_j^2 E\left(\frac{1}{u_j^{2/p}} \right)$$

and

$$E\left(\frac{1}{u_j^{2/p}} \right) = \int_0^\infty e^{-x}(x^{-2/p})dx$$

$$= \int_0^1 e^{-x}(x^{-2/p})dx + \int_1^\infty e^{-x} \cdot (x^{-2/p})dx$$

$$= \int_0^1 x^{-2/p}dx + \int_1^\infty e^{-x}dx. \quad \text{(trivial bounds on } e^{-x} \text{ and } x^{-2/p})$$

The second integral trivially converges, and the former one converges because $p > 2$. This gives that

$$E(\|Q\|^2) = O(\|x\|_2^2)$$

which gives that with high probability we will have $\sigma^2 \leq O(\|x\|_2^2)/m$.

To use Bernstein's inequality, we will associate this bound on σ^2, which is given in terms of $\|x\|_2$, to a bound in terms of $\|x\|_p$. By using an argument based on Hölder's inequality,

Theorem 1.7 (Hölder's inequality) *Let $f, g \in \mathbb{R}^n$. Then*

$$\sum_i f_i g_i \leq \|f\|_a \|g\|_b$$

for any $1 \leq a, b \leq \infty$ satisfying $1/a + 1/b = 1$.

Here $f_i = x_i^2$, $g_i = 1$, $a = p/2$, $b = 1/(1 - a)$ gives

$$\|x_i\|^2 = \sum_i f_i g_i$$

$$\leq \left(\sum_i (x_i^2)^{p/2}\right)^{2/p} \left(\sum_i 1^{1/(1-2/p)}\right)^{1-2/p}$$

$$\leq \|x\|_p^2 \cdot n^{1-2/p}$$

Using the fact that we chose m to $\Theta(n^{1-2/p} \lg(n))$, we can then obtain the following bound on σ^2 with high probability.

$$\sigma^2 \leq O\left(\frac{\|x\|_2^2}{m}\right)$$

$$\leq O\left(\frac{T^2 n^{1-2/p}}{m}\right)$$

$$\leq O\left(\frac{T^2 n^{1-2/p}}{n^{1-2/p} \lg n}\right)$$

$$\leq O\left(\frac{T^2}{\lg(n)}\right)$$

Now let us use Bernstein's inequality to prove the required result.

$$P\left(\left|\sum R_i\right| > T/10\right) \lesssim e^{-cT^2/100 \cdot O(\lg(n)/T^2)} + e^{-cT/10 \cdot (v \lg(n)/T)}$$

$$\leq e^{-F \lg(n)}$$

$$= n^F$$

So the probability that the noise at most $T/10$ can be made poly n. But there are at most n buckets, which means that a union bound gives us that with at least constant probability all of the light index contributions are are at most $T/10$.

Distinct elements are used in SQL to efficiently count distinct entries in some column of a data table. It is also used in network anomaly detection to, track the rate at which a worm is spreading. You run distinct elements on a router to count how many distinct entities are sending packets with the worm signature through your router.

For more general moment estimation, there are other motivating examples as well. Imagine x_i is the number of packets sent to IP address i. Estimating $\|x\|_\infty$ would give an approximation to the highest load experienced by any server. Obviously, as elaborated earlier, $\|x\|_\infty$ is difficult to approximate in small space, so in practice we settle for the closest possible norm to the ∞-norm, which is the 2-norm.

1.9 Heavy Hitters Problem

Data stream algorithms have become an indispensable tool for analysing massive data sets. Such algorithms aim to process huge streams of updates in a single pass and store a compact summary from which properties of the input can be discovered, with strong guarantees on the quality of the result. This approach has found many applications, in large scale data processing and data warehousing, as well as in other areas, such as network measurements, sensor networks and compressed sensing. One high-level application example is computing popular products. For example, A could be all of the page views of products on amazon.com yesterday. The heavy hitters are then the most frequently viewed products.

Given a stream of items with weights attached, find those items with the greatest total weight. This is an intuitive problem, which relates to several natural questions: given a stream of search engine queries, which are the most frequently occurring terms? Given a stream of supermarket transactions and prices, which items have the highest total euro sales? Further, this simple question turns out to be a core subproblem of many more complex computations over data streams, such as estimating the entropy, and clustering geometric data. Therefore, it is of high importance to design efficient algorithms for this problem, and understand the performance of existing ones.

The problem can be solved efficiently if A is promptly obtainable in main memory then simply sort the array and do a linear scan over the result, outputting a value if and only if it occurs at least n/k times. But, what about solving the Heavy Hitters problem with a single pass over the array?

In Point Query, we are given some $x \in \mathbb{R}^n$ updated in a turnstile model, with n large. Suppose that x has a coordinate for each string your search engine could see and x_i is the number of times we have seen string i. We seek a function query(i) that, for $i \in [n]$, returns a value in $x_i \pm \varepsilon \cdot \|x\|_1$.

In Heavy Hitters, we have the same x but we need to compute a set $L \subset [n]$ such that

1. $|x_i| \geq \varepsilon \|x\|_1 \Rightarrow i \in L$
2. $|x_i| < \frac{\varepsilon}{2} \|x\|_1 \Rightarrow i \notin L$

If we can solve Point Query with bounded space then we can solve Heavy Hitters with bounded space as well (but without efficient run-time). So, we just run Point Query with $\varepsilon/10$ on each $i \in [n]$ and output the set of indices i for which we had large estimates of x_i.

Now let us define an *incoherent matrix*.

Definition 1.3 $\Pi \in \mathbb{R}^{m \times n}$ is *ε-incoherent* if

1. For all i, $\|\Pi_i\|_2 = 1$
2. For all $i \neq j$, $|\langle \Pi_i, \Pi_j \rangle| \leq \varepsilon$.

We also define a related object: a *code*.

Definition 1.4 An (ε, t, q, N)-code is a set $\mathscr{F} = \{F_1, \ldots, F_N\} \subseteq [q]^t$ such that for all $i \neq j$, $\Delta(F_i, F_j) \geq (1 - \varepsilon)t$, where Δ indicates Hamming distance.

The key property of a code can be summarized verbally: any two distinct words in the code agree in at most εt entries.

There is a relationship between incoherent matrices and codes.

Claim *Existence of an (ε, t, q, n)-code implies existence of an ε-incoherent Π with $m = qt$.*

Proof We construct Π from $\mathscr{F}$. We have a column of Π for each $F_i \in \mathscr{F}$, and we break each column vector into t blocks, each of size q. Then, the jth block contains binary string of length q whose ath bit is 1 if the jth element of F_i is a and 0 otherwise. Scaling the whole matrix by $1/\sqrt{t}$ gives the desired result.

Claim *Given an ε-incoherent matrix, we can create a linear sketch to solve Point Query.*

Claim *A random code with $q = O(1/\varepsilon)$ and $t = O(\frac{1}{\varepsilon} \log N)$ is an (ε, t, q, N)-code.*

1.10 Count-Min Sketch

Next we will consider another algorithm where the objective is to know the frequency of popular items. The idea is we can hash each incoming item several different ways, and increment a count for that item in a lot of different places, one place for each hash. Since each array that we use is much smaller than the number of unique items that we see, it will be common for more than one item to has to a particular location. The trick is that for the any of most common items, it is very likely that at least one of the hashed locations for that item will only have collisions with less common items. That means that the count in that location will be mostly driven by that item. The problem is how to find the cell that only has collisions with less popular items.

In other words, Count-Min (CM) sketch is a compact summary data structure capable of representing a high-dimensional vector and answering queries on this vector, in particular point queries and dot product queries, with strong accuracy guarantees. Such queries are at the core of many computations, so the structure can be used in order to answer a variety of other queries, such as frequent items (heavy hitters), quantile finding, and join size estimation (Cormode and Muthukrishnan 2005). Since the data structure can easily process updates in the form of additions or subtractions to dimensions of the vector, which may correspond to insertions or deletions, it is capable of working over streams of updates, at high rates. The data structure maintains the linear projection of the vector with a number of other random vectors. These vectors are defined implicitly by simple hash functions. Increasing the range of the hash functions increases the accuracy of the summary, and increasing the number of hash functions decreases the probability of a bad estimate. These tradeoffs are quantified precisely below. Because of this linearity, CM sketches can be scaled, added and subtracted, to produce summaries of the corresponding scaled and combined vectors.

Thus for CM, we have streams of insertions, deletions, and queries of how many times a element could have appeared. If the number is always positive, it is called Turnstile Model. For example, in a music party, you will see lots of people come in and leave, and you want to know what happens inside. But you do not want to store every thing happened inside, you want to store it more efficiently.

One application of CM might be you scanning over a corpus of a lib. There are a bunch of URLs you have seen. There are huge number of URLs. You cannot remember all URLs you see. But you want to estimate the query about how many times you saw the same URLs. What we can do is to store a set of counting bloom filters. Because a URL can appear multiple times, how would you estimate the query given the set of counting bloom filter?

We can take the minimal of all hashed counters to estimate the occurrence of a particular URL. Specifically:

$$\text{query}(x) = \min_i y_{h_i(x)}$$

See that the previous analysis about the overflow of counting bloom filters does work.

Then there is a question of how accurate the query is? Let $F(x)$ be the real count of an individual item x. One simple bound of accuracy can be

$$\forall i, \, e[y_{h_i(x)} - F(x)] = \frac{\big(\|F\|_1 - F(x)\big) \cdot k}{m}$$
$$\leq \frac{\|F\|_2 \ln 2}{n}$$

which tells us the average error for all single hashed places with regard to the real occurrence. So we know that it is always overestimated.

For the total number of items $\sum_x F(x)$, we have ($F(x)$ is non-negative)

$$\sum_x F(x) = \|F\|_1$$

where, in general, $\forall z \in \mathbb{R}^d, \|z\|_1 = \sum_{i=1}^d |z_i|, \|z\|_2 = \sum_{i=1}^d z_i^2, \|z\|_\infty = \max_{i \in [1,d]} |z_i|$.

See that you do not even need m to be larger than n. If you have a huge number of items, you can choose m to be very small (m can be millions for billions of URLs).

Now we have bound of occurrence estimation for each individual i in expectation. However, what we really need to concern is the query result. We know that

$$\forall i, k, \; y_{h_i(x)} - F(x) \leq 2\|F\|_1 \frac{k}{m} \quad \text{w.p. } \frac{1}{2}$$

And now if I choose $k = \log \frac{1}{\delta}$,

$$\min y_{h_i(x)} - F(x) \leq 2\|F\|_1 \frac{k}{m} \quad \text{w.p. } 1 - \delta$$

If $F(x)$ is concentrated in a few elements, the tth largest is proportional to roughly $\frac{1}{t^\alpha}$ with the power law distribution. So if we choose m to be small, then you can estimate the top URLs pretty well.

$$\sum F(x) = \frac{m^2}{k} = \mathcal{O}(1)$$

In fact, you can show a better result for CM, which is rather than having your norm depend on the 1-norm. There could be a few elements having all of occurrences. For example, several people have been visiting google.com. The top few URLs have almost all the occurrences. Then probably for a given URL, it might collide some of them in some of the time. But probably one of them is not going to collide, and probably most of them are going to collide. So one can get in terms of l-1 norm but in terms of l-1 after dropping the top k elements. So given billions of URLs, you can drop the top ones and get l-1 norm for the residual URLs.

$$\sum_{\text{non-top } k} F(x) \approx \frac{1}{m/k} = \frac{k}{m}$$

The Count-Min sketch has found a number of applications. For example, Indyk (Indyk 2003) used the Count-Min Sketch to estimate the residual mass after removing a set of items. This supports clustering over streaming data. Sarlós et al. (Sarlós et al. 2006) gave approximate algorithms for personalized page rank computations which make use of Count-Min Sketches to compactly represent web-size graphs.

1.10.1 Count Sketch

One of the important fundamental problems on a data stream is that of finding the most frequently occurring items in the stream. We shall assume that the stream is large enough that memory-intensive solutions such as sorting the stream or keeping a counter for each distinct element are infeasible, and that we can only afford to process the data by making one or more passes over it. This problem arises in the context of search engines, where the streams in question are streams of queries sent to the search engine and we are interested in finding the most frequent queries handled in some period of time. Interestingly, in the context of search engine query streams, since the queries whose frequency changes most between two consecutive time periods can indicate which topics are increasing or decreasing in popularity at the fastest rate. Reference (Charikar et al. 2002) presented a simple data structure called a count-sketch and developed a 1-pass algorithm for computing the count-sketch of a stream. Using a count sketch, one can consistently estimate the frequencies of the most common items. Reference (Charikar et al. 2002) showed that the count-sketch data structure is additive, i.e. the sketches for two streams can be directly added or subtracted. Thus, given two streams, we can compute the difference of their sketches, which leads to a 2-pass algorithm for computing the items whose frequency changes the most between the streams.

The Count Sketch (Charikar et al. 2002) is basically like CM, except that when you do hashing, you also associate the sum with each hash function h.

$$y_j = \sum_{(i,x):h_i(x)=j} F(x)S_i(x)$$

Then the query can be defined as

$$\text{query}(x) = \text{median } S_i(x)y_{h_i(x)}$$

The error can be converted from l-1 norm to l-2 norm.

$$error^2 \lesssim \frac{||F_{\frac{m}{k}}||_2^2}{\frac{m}{k}}$$

On top of that, suppose everything else is 0, then $y_{h_i(x)} \approx S_i(x)F(x)$. So we will have

$$\text{query}(x) \approx \text{median } \left(S_i(x)\right)^2 F(x)$$

Then if there is nothing special going on, the query result would be $F(x)$.

1.10.2 Count-Min Sketch and Heavy Hitters Problem

The Count-Min (CM) Sketch is an example of a sketch that permits a number of related quantities to be estimated with accuracy guarantees, including point queries and dot product queries. Such queries are very crucial for several computations, so the structure can be used in order to answer a variety of other queries, such as frequent items (heavy hitters), quantile finding, join size estimation, and so on. Let us consider the CM sketch, that can be used to solve the ε-approximate heavy hitters (HH) problem. It has been implemented in real systems. A predecessor of the CM sketch (i.e. count sketch) has been implemented on top of their MapReduce parallel processing infrastructure at Google. The data structure used for this is based on hashing.

Definition 1.5 The Count-Min (CM) sketch (Cormode and Muthukrishnan 2005)

1. Hashing $h_1, \ldots h_L : [n] \to [t]$
2. counters $F_{a,b}$ for $a \in [L], b \in [t]$
3. $F_{a,b} = \sum_{i \in [n], h_a(i)=b} x_i$
4. for ε-point query with failure probability δ, set $t = 2/\varepsilon, L = \lg(1/\delta)$.
   And let $query(i)$ output $\min_{i \le r \le L} F_{r,h_r(i)}$ (assuming "strict turnstile", for any i, $x_i \ge 0$).

Claim $query(i) = x_i \pm \varepsilon\|x\|_1$ w.p $\ge 1 - \delta$. $m = O(\varepsilon^{-1}\lg(1/\delta))$.

Proof CM sketch

1. Fix i, let $Q_j = 1$ if $h_r(j) = h_r(i)$, $Q_j = 0$ otherwise. $F_{r,h_r(i)} = x_i + \sum_{j \ne i} x_j Q_j$ error E.
2. We have $\mathbb{E}(E) = \sum_{j \ne i} |x_j| \mathbb{E} Q_j = \sum_{j \ne i} |x_j|/t \le \varepsilon/2 \cdot \|x\|_1$
3. $\mathbb{P}(E > \varepsilon\|x\|_1) < 1/2$
4. $\mathbb{P}(\min_r F_{r,h_r(i)} > x_i + \varepsilon\|x\|_1) < 1/2^E = \delta$

Theorem 1.8 *There is an α-Heavy Hitter (strict turnstile) w.p $1 - \eta$.*

Proof We can perform point query with $\varepsilon = \alpha/4, \delta = \eta/n \to m = O(1/\alpha \log(n/\eta))$ with query time $O(n \cdot \log(n/\eta))$.

Interestingly, a binary tree using n vector elements as the leaves can be illustrate as follows:

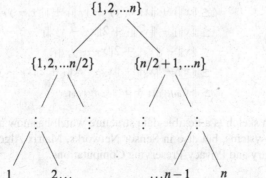

The above tree has $\lg n$ levels and the weight of each vertex is the sum of elements. Here we can utilise a $CountMin$ algorithm for each level.

The procedure:

1. Run Count-Min from the roots downward with error $\varepsilon = \alpha/4$ and $\delta = \eta\alpha/4\log n$
2. Move down the tree starting from the root. For each vertex, run CountMin for each of its two children. If a child is a heavy hitter, i.e. CountMin returns $\ge 3\alpha/4\|x\|_1$, continue moving down that branch of the tree.
3. Add to L any leaf of the tree that you point query and that has $CM(i) \ge 3\alpha/4\|x\|_1$.

The l_1 norm will be the same at every level since the weight of the parents vertex is exactly the sum of children vertices. Next vertex u contains heavy hitter amongst leaves in its subtree $\rightarrow$ u is hit at its level. There is at most $2/\alpha$ vertices at any given level which are $\alpha/2$-heavy hitter at that level. This means that if all point queries correct, we only touch at most $(2/\alpha)\lg n$ vertices during Best First Search. For each CM_j, we have $\varepsilon = \alpha/4, \delta = \eta\alpha/4\log n \rightarrow space(CM_j) = O(1/\alpha \cdot \log(\log n/\alpha\eta)) \rightarrow totalSpace = O(1/\alpha \cdot \log n \cdot \log(\log n/\alpha\eta))$.

We know heavy hitter is l_∞/l_1 guarantee. To be precise $\|x - x'\|_1 \leq (1 + \varepsilon)\|x_{tail(k)}\|_1$. You can get to l_1/l_1 for Heavy Hitters and CM sketch can give it with $\|x'\|_0 \leq k$.

Definition 1.6 $x_{tail(k)}$ is x with the heaviest k coordinates in magnitude reduced to zero.

Claim If CM has $t \geq \Theta(k/\varepsilon)$, $L = \Theta(\lg(1/\delta))$ then w.p. $1 - \delta$, $x_i' = x_i \pm \varepsilon/k\|x_{tail(k)}\|_1$.

Given x' from CM output $(x_i' = query(i))$. Let $T \subset [n]$ correspond to largest k entries of x' in magnitude. Now consider $y = x_T'$.

Claim $\|x - y\|_1 \leq (1 + 3\varepsilon)\|x_{tail(k)}\|_1$.

Proof Let S denote $head(x) \subset [n]$ and T denote $head(x') \subset [n]$. We have

$$
\begin{aligned}
\|x - y\|_1 &= \|x\|_1 - \|x_T\|_1 + \|x_T - y_T\|_1 \\
&\leq \|x\|_1 + \|x_T - y_T + y_T\|_1 + \|x_T - y_T\|_1 \\
&\leq \|x\|_1 - \|y_T\|_1 + 2\|x_T - y_T\|_1 \\
&\leq \|x\| - \|y_S\| + 2\|x_T - y_T\|_1 \\
&\leq \|x\| - \|x_S\| + \|x_S - y_S\|_1 + 2\|x_T - y_T\|_1 \\
&\leq \|x_{tail(k)}\|_1 + 3\varepsilon\|x_{tail(k)}\|_1
\end{aligned}
$$

Count-Min sketch is a flexible data structure which has now applications within Data Stream systems, but also in Sensor Networks, Matrix Algorithms, Computational Geometry and Privacy-Preserving Computations.

1.11 Streaming k-Means

The aim is to design light-weight algorithms that make only one pass over the data. Clustering techniques are largely used in machine learning applications, as a way to summarise large quantities of high-dimensional data, by partitioning them into clusters that are useful for the specific application. The problem with many heuristics designed to implement some notion of clustering is that their outputs can be hard to evaluate. Approximation guarantees, with respect to some valid objective, are thus useful. The k-means objective is a simple, intuitive, and widely-used clustering for data in Euclidean space. However, although many clustering algorithms have been designed with the k-means objective in mind, very few have approximation guarantees with respect to this objective. The problem to solve is that k-means clustering requires multiple tries to get a good clustering and each try involves going through the input data several times.

This algorithm will do what is normally a multi-pass algorithm in exactly one pass. In general, problem in k-means is that you wind up with clusterings containing bad initial conditions. So, you will split some clusters and other clusters will be joined together as one. Therefore you need to restart k-means. k-means is not only multi-pass, but you often have to carry out restarts and run it again. In case of multi-dimensional complex data ultimately you will get bad results.

But if we could come up with a small representation of the data, a sketch, that would prevent such problem. We could do the clustering on the sketch instead on the data. Suppose if we can create the sketch in a single fast pass through the data, we have effectively converted k-means into a single pass algorithm. The clustering with too many clusters is the idea behind streaming k-means sketch. All of the actual clusters in the original data have several sketch centroids in them, and that means, you will have something in every interesting feature of the data, so you can cluster the sketch instead of the data. The sketch can represent all kinds of impressive distributions if you have enough clusters. So any kind of clustering you would like to do on the original data can be done on the sketch.

1.12 Graph Sketching

Several kinds of highly structured data are represented as graphs. Enormous graphs arise in any application where there is data about both basic entities and the relationships between these entities, e.g., web-pages and hyperlinks; IP addresses and network flows; neurons and synapses; people and their friendships. Graphs have also become the *de facto* standard for representing many types of highly-structured data. However, analysing these graphs via classical algorithms can be challenging given the sheer size of the graphs (Guha and McGregor 2012).

A simple approach to deal with such graphs is to process them in the data stream model where the input is defined by a stream of data. For example, the stream could consist of the edges of the graph. Algorithms in this model must process the input stream in the order it arrives while using only a limited amount memory. These constraints capture different challenges that arise when processing massive data sets, e.g., monitoring network traffic in real time or ensuring I/O efficiency when processing data that does not fit in main memory. Immediate question is how to trade-off size and accuracy when constructing data summaries and how to quickly update these summaries. Techniques that have been developed to the reduce the space use have also been useful in reducing communication in distributed systems. The model also has deep connections with a variety of areas in theoretical computer science including communication complexity, metric embeddings, compressed sensing, and approximation algorithms.

Traditional algorithms for analyzing properties of a graph are not appropriate for massive graphs because of memory constraints. Often the graph itself is too large to be stored in memory on a single computer. There is a need for new techniques, new algorithms to solve graph problems such as, checking if a massive graph is connected, if it is bipartite, if it is k-connected, approximating the weight of a minimum spanning tree. Moreover, storing a massive graph requires usually $O(n^2)$ memory, since that is the maximum number of edges the graph may have. In order to avoid using that much memory and one can make a constraint. The *semi-streaming model* is a widely used model of computation which restricts to using only $O(n\,poly\log n)$ memory, where *polylog n* is a notation for a polynomial in $\log n$.

When processing big data sets, a core task is to construct synopses of the data. To be useful, a synopsis data structure should be easy to construct while also yielding good approximations of the relevant properties of the data set. An important class of synopses are sketches based on linear projections of the data. These are applicable in many models including various parallel, stream, and compressed sensing settings.

We discuss graph sketching where the graphs of interest encode the relationships between these entities. Sketching is connected to dimensionality reduction. The main challenge is to capture this richer structure and build the necessary synopses with only linear measurements.

Let $G = (V, E)$, where we see edges $e \in E$ in stream. Let $|V| = n$ and $|E| = m$. We begin by providing some useful definitions:

Definition 1.7 A graph is bipartite if we can divide its vertices into two sets such that: any edge lies between vertices in opposite sets.

Definition 1.8 A cut in a graph is a partition of the vertices into two disjoints sets. The cut size is the number of edges with endpoints in opposite sets of the partition.

Definition 1.9 A minimum spanning tree (MST) is a tree subgraph of the input graph that connects all vertices and has minimum weight among all spanning trees.

Given a connected, weighted, undirected graph $G(V, E)$, for each edge $(u, v) \in E$, there is a weight $w(u, v)$ associated with it. The Minimum Spanning Tree (MST) problem in G is to find a spanning tree $T(V, E')$ such that the weighted sum of the edges in T is minimized, i.e.

$$\text{minimize } w(T) = \sum_{(u,v)\in E'} w(u, v), \text{ where } E' \subseteq E$$

For instance, the diagram below shows a graph, G, of nine vertices and 12 weighted edges. The bold edges form the edges of the MST, T. Adding up the weights of the MST edges, we get $w(T) = 140$.

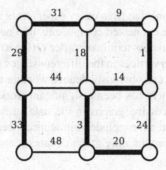

Definition 1.10 The order of a graph is the number of its vertices.

Claim *Any deterministic algorithm needs $\Omega(n)$ space.*

Proof Suppose we have $x \in \{0, 1\}^{n-1}$. As before, we will perform an encoding argument. We create a graph with n vertices $0, 1, \ldots, n-1$. The only edges that exist are as follows: for each i such that $x_i = 1$, we create an edge from vertex 0 to vertex i. The encoding of x is then the space contents of the connectivity streaming algorithm run on the edges of this graph. Then in decoding, by querying connectivity between 0 and i for each i, we can determine whether x_i is 1 or 0. Thus the space of the algorithm must be at least $n-1$, the minimum encoding length for compressing $\{0, 1\}^{n-1}$.

For several graph problems, it turns out that $\Omega(n)$ space is required. This motivated the Semi-streaming model for graphs (Feigenbaum et al. 2005), where the goal is to achieve $O(n \lg^c n)$ space.

1.12.1 Graph Connectivity

Consider a dynamic graph stream in which the goal is to compute the number of connected components using memory limited to $O(n \text{ polylog } n)$. The idea is to use a basic algorithm and reproduce it using sketches. See the following algorithm.

Algorithm
Step 1: For each vertex pick an edge that connects it to a neighbour.
Step 2: Contract the picked edges.
Step 3: Repeat until there are no more edges to pick in step 1.
Result: the number of connected components is the number of vertices at the end of the algorithm.

Finally, consider a non-sketch procedure, which is based on the simple $O(\log n)$ stage process. In the first stage, we find an arbitrary incident edge for each vertex. We then collapse each of the resulting connected components into a supervertex. In each subsequent stage, we find an edge from every supervertex to another supervertex, if one exists, and collapse the connected components into new supervertices. It is not difficult to argue that this process terminates after $O(\log n)$ stages and that the set of edges used to connect supervertices in the different stages include a spanning forest of the graph. From this we can obviously deduce whether the graph is connected.

In the past few years, there has been a significant work on the design and analysis of algorithms for processing graphs in the data stream model. Problems that have received substantial attention include estimating connectivity properties, finding approximate matching, approximating graph distances, and counting the frequency of sub-graphs.

Chapter 2
Sub-linear Time Models

2.1 Introduction

Sub-linear time algorithms represent a new paradigm in computing, where an algorithm must give some sort of an answer after inspecting only a very small portion of the input. It has its roots in the study of massive data sets that occur more and more frequently in various applications. Financial transactions with billions of input data and Internet traffic analyses are examples of modern data sets that show unprecedented scale. Managing and analysing such data sets forces us to reconsider the old idea of efficient algorithms. Processing such massive data sets in more than linear time is by far too expensive and frequently linear time algorithms may be extremely slow. Thus, there is the need to construct algorithms whose running times are not only polynomial, but rather are sub-linear in n.

In this chapter we study sub-linear time algorithms which are aimed at helping us understand massive datasets. The algorithms we study inspect only a tiny portion of an unknown object and have the aim of coming up with some useful information about the object. Algorithms of this sort provide a foundation for principled analysis of truly massive data sets and data objects.

The aim of algorithmic research is to design efficient algorithms, where efficiency is typically measured as a function of the length of the input. For instance, the elementary school algorithm for multiplying two n digit integers takes roughly n^2 steps, while more sophisticated algorithms have been devised which run in less than $nlog^2n$ steps. It is still not known whether a linear time algorithm is achievable for integer multiplication. Obviously any algorithm for this task, as for any other non-trivial task, would need to take at least linear time in n, since this is what it would take to read the entire input and write the output. Thus, showing the existence of a linear time algorithm for a problem was traditionally considered to be the gold standard of achievement. Analogous to the reasoning that we used for multiplication, for most natural problems, an algorithm which runs in sub-linear time must necessarily use randomization and must give an answer which is in some sense imprecise. Nevertheless, there are many situations in which a fast approximate solution is more useful than a slower exact solution.

Constructing a sub-linear time algorithm may seem to be an extremely difficult task since it allows one to read only a small fraction of the input. But, in last decade, we have seen development of sub-linear time algorithms for optimization problems arising in such diverse areas as graph theory, geometry, algebraic computations, and computer graphics. The main research focus has been on designing efficient algorithms in the framework of property testing, which is an alternative notion of approximation for decision problems. However, more recently, we see some major progress in sub-linear-time algorithms in the classical model of randomized and approximation algorithms.

Let us begin by proving space lower bounds. The problems we are going to look at are F_0 (distinct elements)-specifically any algorithm that solves F_0 within a factor of ε must use $\Omega(1/\varepsilon^2 + \log n)$ bits. We're also going to discuss **median**, or randomized exact median, which requires $\Omega(n)$ space. Finally, we'll see F_p or $\|x\|_p$, which requires $\Omega(n^{1-2/p})$ space for a 2-approximation.

Suppose we have Alice and Bob, and a function $f : X \times Y \to \{0, 1\}$. Alice gets $x \in X$, and Bob gets $y \in Y$. They want to compute $f(x, y)$. Suppose that Alice starts the conversation. Suppose she sends a message m_1 to Bob. Then Bob replies with m_2, and so on. After k iterations, someone can say that $f(x, y)$ is determined. The aim for us is to minimize the total amount of communication, or $\sum_{i=1}^{k} |m_i|$, where the absolute value here refers to the length of the binary string.

One of the application domains for communication complexity is distributed computing. When we wish to study the cost of computing in a network spanning multiple cores or physical machines, it is very useful to understand how much communication is necessary, since communication between machines often dominates the cost of the computation. Accordingly, lower bounds in communication complexity have been used to obtain many negative results in distributed computing. All applications of communication complexity lower bounds in distributed computing to date have used only two-player lower bounds. The reason for this appears to be twofold: First, the models of multi-party communication favoured by the communication complexity community, the number-on-forehead model and the number in-hand broadcast model, do not correspond to most natural models of distributed computing. Second, two-party lower bounds are surprisingly powerful, even for networks with many players. A typical reduction from a two-player communication complexity problem to a distributed problem T finds a sparse cut in the network, and shows that, to solve T, the two sides of the cut must implicitly solve, say, set disjointness.

A *communication protocol* is a manner in which discourse agreed upon ahead of time, where Alice and Bob both know f. There's obvious the two obvious protocols, where Alice sends $\log X$ bits to send x, or where Bob sends y via $\log Y$ bits to Alice. The aim is to either beat these trivial protocols or prove that none exists.

There is a usual connection between communication complexity and space lower bounds as follows: a communication complexity lower bound can yield a streaming lower bound. We'll restrict our attention to 1-way protocols, where Alice just sends messages to Bob. Suppose that we had a lower bound for a communication problem-

Alice has $x \in X$, and Bob has $y \in Y$ and we know that the lower bound on the optimal communication complexity is $\overrightarrow{D}(f)$. The D here refers to the fact that the communication protocol is deterministic. In case of a streaming problem, Alice can run her streaming algorithm on x, the first half of the stream, and send the memory contents across to Bob. Bob can then load it and pass y, the second half of the stream, and calculate $f(x, y)$, the ultimate result. Hence the minimal amount of space necessary is $\overrightarrow{D}(f)$.

Exact and deterministic F_0 requires $\Omega(n)$ space. We will use a reduction, because the F_0 problem must be hard, otherwise we could use the above argument. We use the *equality problem* (EQ), which is where $f(x, y) = x == y$. We claim $D(EQ) = \omega(n)$. This is straightforward to prove in the one-way protocol, by using the pigeonhole principle (Nelson 2015).

In order to reduce EQ to F_0 let us suppose that there exists a streaming algorithm A for F_0 that uses S bits of space. Alice is going to run A on her stream x, and then send the memory contents to Bob. Bob then queries F_0, and then for each $i \in y$, he can append and query as before, and solve the equality problem. Nonetheless, this solves EQ, which requires $\Omega(n)$ space, so S must be $\Omega(n)$.

Let us define,

- $D(f)$ is the optimal cost of a deterministic protocol.
- $R_\delta^{pub}(f)$ is the optimal cost of the random protocol with failure probability δ such that there is a shared random string (written in the sky or something).
- $R_\delta^{pri}(f)$ is the same as above, but each of Alice/Bob have private random strings.
- $D_{\mu,s}(f)$ is the optimal cost of a deterministic protocol with failure probability δ where $(x, y) \sim \mu$.

Claim $D(f) \geq R_\delta^{pri}(f) \geq R_\delta^{pub}(f) \geq D_{\mu,s}(f)$.

Proof The first inequality is obvious, since we can simulate the problem. The second inequality follows from the following scheme: Alice just uses the odd bits, and Bob just uses the even bits in the sky. The final inequality follows from an indexing argument: suppose that P is a public random protocol with a random string s, $\forall(x, y)\mathbb{P}(P_s\text{correct}) \geq 1 - \delta$. Then there exists an s^* such that the probability of P_s succeeding is large. See that s^* depends on μ.

If we want to have a lower bound on deterministic algorithms, we need to lower bound $D(f)$. If we want to have the lower bound of a randomized algorithm, we need to lower bound $R_\delta^{pri}(f)$. We need Alice to communicate the random bits over to Bob so that he can keep on running the algorithm, and we need to *include* these bits in the cost since we store the bits in memory. So, to lower bound randomized algorithms, we lower bound $D_{\mu,s}(f)$.

Interestingly one can solve EQ using public randomness with constant number of bits. If we want to solve it using private randomness for EQ, we need $\log n$ bits. Alice picks a random prime, and she sends $x \mod p$ and sends across $x \mod p$ and the prime. Neumann's theorem says that you can reverse the middle inequality in the above at a cost of $\log n$.

2.2 Fano's Inequality

Fano's inequality is a well-known information-theoretical result that provides a lower bound on worst-case error probabilities in multiple-hypotheses testing problems. It has important consequences in information theory and related fields. In statistics, it has become a major tool to derive lower bounds on minimax (worst-case) rates of convergence for various statistical problems such as nonparametric density estimation, regression, and classification.

Suppose you need to make some decision, and I give you some information that helps you to decide. Fano's inequality gives a lower bound on the probability that you end up making the wrong choice as a function of your initial uncertainty and how newsy my information was. Interestingly, it does not place any constraint on how you make your decision. i.e., it gives a lower bound on your best case error probability. If the bound is negative, then in principle you might be able to eliminate your decision error. If the bound is positive (i.e., binds), then there is no way for you to use the information I gave you to always make the right decision.

Now let us consider a two-player problem: *INDEX*. Alice gets $x \in \{0, 1\}^n$, and Bob gets $j \in [n]$, and $INDEX(x, j) = x_j$.

We are going to show that *INDEX*, the problem of finding the jth element of a streamed vector, is hard. Then, we'll show that this reduces to *GAPHAM*, or Gap Hamming which will reduce to F_0. Also, INDEX reduces to *MEDIAN*. Finally, DISJ$_t$ reduces (with $t = (2n)^{1/p}$) to F_p, $p > 2$.

Claim $R_\delta^{pub\to}(INDEX) \geq (1 - \mathfrak{H}_2(\delta))n$,   where $\mathfrak{H}_2(\delta) = \delta \log(\delta) + (1 - \delta)$ $\log(1 - \delta)$, *the entropy function. If $\delta \approx 1/3$.*

In fact, the distributional complexity has the same lower bound.

Let us first elaborate some definitions.

Definitions: if we have a random variable X, then

- $H(X) = \sum_x p_x \log(p_x)$ *(entropy)*
- $H(X, Y) = \sum_{(x,y)} p_{x,y} \log p_{x,y}$ *(joint entropy)*
- $H(X|Y) = e_y(H(X|Y = y))$ *(conditional entropy)*
- $I(X, Y) = H(X) - H(X|Y)$ *(mutual information)*

The *entropy* is the amount of information or bits we need to send to communicate $x \in X$ in expectation. This can be achieved via Huffman coding (in the limit). The mutual information is how much of X we get by communicating Y.

The following are some fundamental rules considering these equalities

Lemma 2.1

- *Chain rule:* $H(X, Y) = H(X) + H(Y|X)$.
- *Chain rule for mutual information:* $I(X, Y|Q) = I(X, Q) + I(Y, Q|X)$.
- *Subadditivity:* $H(X, Y) \leq H(X) + H(Y)$.
- *Chain rule + subadditivity:* $H(X|Y) \leq H(X)$.
- *Basic* $H(X) \leq \log |supp(X)|$.
- $H(f(X)) \leq H(X) \ \forall f$.

Theorem 2.1 (Fano's Inequality) *If there exist two random variables X, Y and a predictor g such that $\mathbb{P}(g(Y) \neq X) \leq \delta)$, then $H(X|Y) \leq H_2(\delta) + \delta \cdot \log_2(|supp(X)| - 1)$.*

If X is a binary random variable then the second term vanishes. If all you have is Y, based on Y you make a guess of X. Ultimately for small δ, $H_2(\delta) \approx \delta$.

Let Π be the transcript of the optimal communication protocol. It's a one-way protocol. So, we know that $R_\delta^{pub}(INDEX) \geq H(\Pi) \geq I(\Pi, \text{input}) = I(\Pi, \text{input})$.

We know that for all x and for all j, $\mathbb{P}_s(\text{Bob is correct}) \geq 1 - \delta$, which implies that for all j, $\mathbb{P}_{X \sim Unif} \mathbb{P}_s(\text{Bob is correct}) \geq 1 - \delta$, which them implies that by Fano's inequality,

$$H(X_j | \Pi) \geq H_2(\delta)$$

See that Π is a random variable because of the random string in the sky, and also because it is dependent on X (Nelson 2015).

See that we have

$$|\Pi| \geq I(X; \Pi)$$
$$= \sum_{i=1}^{n} I(X_i; \Pi | X_1, \ldots, X_{i-1}) \text{ (chain rule n times)}$$
$$= \sum_i H(X_i | X^{<i}) - H(X_i | \Pi, X^{<i})$$
$$\geq \sum_i 1 - H_2(\delta) = n(1 - H_2(\delta))$$

Since we have INDEX, let's use it to prove lower bound, namely MEDIAN. We want a randomized, exact median of $x_1, \ldots, x_n$ with probability $1 - \delta$. We shall use a reduction.

Claim *INDEX on $\{0, 1\}^n$ reduces to MEDIAN with $m = 2n + 2$, with string length $2n - 1$. To solve INDEX, Alice inserts $2 + x_1, 4 + x_2, 6 + x_3 \ldots$ into the stream, and Bob inserts $n - j$ copies of 0, and another $j - 1$ copies of $2n + 2$.*

Suppose that $n = 3$ and $x = 101_2$. Then Alice will choose 3, 4, 7 out of 2, 3, 4, 5, 6, 7. Bob cares about a particular index, suppose the first index. Bob is going to make this stream length 5, such that the median of the stream is exactly the index he wants. Basically, we can insert 0 or $2n + 2$ exactly where we want, moving around the j index to be the middle, which then we can then output.

Now we use INDEX to give a lower bound on the space usage of randomized exact F_0. We then present a lower bound for randomized approximate F_0, something that we have thus far been unable to do. We then provide a lower bound on F_p via the disjointness problem. Then we move on to dimensionality reduction, distortion, and distributional Johnson–Lindenstrauss and the fact that it implies Johnson–Lindenstrauss.

2.3   Randomized Exact and Approximate Bound F_0

We show that randomized exact F_0 requires $\Omega(n)$ space with failure probability $\frac{1}{3}$.

Proof We perform the following reduction from INDEX. Let Alice receive $x \in \{0, 1\}^n$ and Bob receive $j \in [n]$. It is then Bob's job to find the j'th index of x. They proceed in the following manner. Alice runs our F_0 algorithm on x and sends both the memory contents of the algorithm and the support of x to Bob. Bob then appends j to the stream and queries F_0 from the the memory contents of the algorithm. If F_0 increases, Bob outputs 0, else he outputs 1. We conclude that for S equal to the space usage of the algorithm

$$S + \log n \geq c * n \implies S = \Omega(n)$$

Where $\log n$ factor comes from sending the support of x.

To show randomized approximate F_0 has space lower bound $\Omega(\log n)$ we first state this theorem that can be found in Kushilevitz and Nisan (Kushilevitz and Nisan 1997). In general, it lower bounds the private communication bound by the log of the deterministic communication bound.

Theorem 2.2 $\forall f : \{0, 1\}^n x \{0, 1\}^n \to \{0, 1\}$, *where f is a communication problem, than*

$$R_{\frac{1}{3}}^{pri} \geq \Omega(\log(D(f)))$$

Proof If we view f as a two player game between Alice and Bob on a binary tree of height s and total leaves 2^s, than Alice and Bob could deterministically simulate the private randomized procedure on this tree. For instance, for any path from root to leaf, Alice can compute the probability she would stay on the path given that Bob does as well. She can then send these probabilities to Bob for every single leaf. Bob can then compute the probabilities he stays on the same paths and can output the final result accordingly.

Now we prove that randomized approximate F_0 has space lower bound $\Omega(\log n)$.

Proof Let C be a subset of $\{0, 1\}^n$ such that $\forall c \in C$ the support of c is $\frac{n}{100}$. Also $\forall c \neq c' \in C$, we have $|c \cap c'| \leq \frac{n}{2000}$. Finally $|C| \geq 2^{\Omega(n)}$. In essence, it is a collection of subsets that are largely disjoint but very numerous. We know deterministic equality, EQ, on C requires $\Omega(n)$ communication. Then using Kushilevitz Nisan we have

$$\implies R_{\frac{1}{3}}^{pri}(EQ_C) \geq \Omega(\log n)$$

Now we notice there is a natural reduction from EQ_C to randomized approximate F_0. Namely, Alice runs F_0 on her set c and sends the memory contents to Bob. Bob then runs c' on the memory contents and determines whether the output for F_0 has roughly doubled. If it has, then $c \neq c'$, if not, than $c = c'$.

2.4 t-Player Disjointness Problem

Let us consider the t-player disjointness problem, for proving lower bounds for the F_p. We have t-players $p_1, p_2, \ldots, p_t$. We assign an n bit string $x_i \in \{0, 1\}^n$ to player p_i. We are then promised that either of the following conditions hold.

1. $\forall i \neq j$ we have $x_i \cap x_j \neq \emptyset$
2. $\exists k \in [n]$ such that $\forall i \neq j$ we have $x_i \cap x_j = \{k\}$

The problem is then to find k with the least communication possible where communication occurs from player 1 passing on to player 2 and so on and so forth until player player t gives the final result.

The proof also uses an information theoretic approach, known as *information complexity* (Chakrabarti et al. 2001). The idea is the following chain of inequalities, where Π is the optimal δ-error communication protocol for some function f: $R_\delta^{pub}(f) = |\Pi| \geq H(\Pi(\mathbf{X})) \geq I(\mathbf{X}; \Pi(X))$, where $\mathbf{X}$ is the set of inputs given to the t players, and $\Pi(\mathbf{X})$ is the transcript of the communication protocol (or the "communication log") when the input is $\mathbf{X}$ (see that it is a random variable since Π uses randomness). Then we define the *information complexity* $IC_{\mu,\delta}(f)$ as the minimum value $I(\mathbf{X}, \Pi(X))$ achievable by any δ-error protocol Π when $\mathbf{X}$ is drawn from distribution μ. Then we have that $R_\delta^{pub}(f) \geq IC_{\mu,\delta}(f)$ for all μ. A variant of this approach was used by (Bar-Yossef et al. 2004) to obtain lower bounds for t-player disjointness, with improvements in (Chakrabarti et al. 2003). The sharp bound was shown in (Gronemeier 2009), with a later work showing how the arguments in (Bar-Yossef et al. 2004) could be strengthened to also get the sharp bound (Jayram 2009).

Theorem 2.3 $R_{\frac{1}{3}}^{pub}(DISJ_t) = \Omega(\frac{n}{t})$.

Remark 2.1 Whilst we do not show that the theorem holds we know that it implies some player sends $\Omega(\frac{n}{t^2})$ bits which we'll demonstrate the following claim.

Claim For $p > 2$ the randomized 1.1 approximation to F_p needs $\Omega(n^{1-\frac{2}{p}})$ bits of space.

Proof Consider $t = \lceil (2n)^{\frac{1}{p}} \rceil$ for the disjoint players problem. Each player generates a virtual stream containing j if and only if $j \in x_i$. Further, we compute F_p on these virtual streams. If all x_i are disjoint then $F_p \leq n$. Alternatively, $F_p \geq t^p \geq 2n$ because some element k must appear at least t times. Since the F_p algorithm is a 1.1 approximation, we can perceive between the two cases. This implies the space usage of the algorithm, S gives

$$S \geq \frac{n}{t^2} = \Omega(n^{1-\frac{2}{p}})$$

hence proved.

2.5 Dimensionality Reduction

Dimensionality reduction (Globerson et al. 2003) has been one of the key techniques used to facilitate the processing of streaming data. The dimension reduction algorithms are generally classified into feature selection, feature extraction and random projection. In simple words, dimension reduction refers to the process of converting a set of data having vast dimensions into data with lesser dimensions ensuring that it conveys similar information concisely. These techniques are typically used while solving machine learning problems to obtain better features for a classification or regression task. The characters of streaming data require the dimension reduction techniques to be as efficient as possible. Thus the common used dimension reduction algorithms used for data streams are random projection and feature selection.

Dimensionality reduction transforms high-dimensional data into lower-dimensional version, such that for the computational problem you are considering, once you solve the problem on the lower-dimensional transformed data, you can get approximate solution on original data. Since the data is in low dimension, your algorithm can run faster.

One of the most used dimensionality reduction techniques is Principal Component Analysis (PCA). The key idea is to find a new coordinate system in which the input data can be expressed with many less variables without a significant error. This new basis can be global or local and can fulfil very different properties. The big data together with the advanced computational resources have attracted the attention of many researchers in Statistics, Computer Science and Applied Mathematics who have developed a wide range of computational techniques dealing with the dimensionality reduction problem.

Dimensionality reduction techniques can be used in different ways including:

1. *Data dimensionality reduction*: Construct a compact low-dimensional encoding of a given high-dimensional data set.
2. *Data visualization*: Give an interpretation of a given data set in terms of intrinsic degree of freedom, usually as a by-product of data dimensionality reduction.
3. *Preprocessing for supervised learning*: Simplify, reduce, and clean the data for subsequent supervised training.

In many large-scale data processing applications, local distances convey more useful information than large distances and are sufficient for uncovering low-dimensional structure. Moreover, there are a variety of situations that rely only on local distances, including nearest-neighbor search, the computation of vector quantization rate-distortion curves, and popular data-segmentation and clustering algorithms. In these cases, it is often desirable to reduce the dimension of the data set for reductions of storage requirements or algorithm running times. If the long distances are unimportant, we may be able to reduce the dimensionality only preserving the local information, and such reduction can be into a far lower dimension than what is possible when attempting to preserve distances between all pairs of points.

Several algorithms for dimensionality reduction have been developed.

Definition 2.1 Suppose we have two metric spaces, (X, d_X), and (Y, d_Y), and a function $f : X \to Y$. Then f has distortion D_f if $\forall x, x' \in X$, $C_1 \cdot d_X(x, x') \leq d_Y(f(x), f(x')) \leq C_2 \cdot d_X(x, x')$, where $\frac{C_2}{C_1} = D_f$.

We will focus on spaces in which $d_X(x, x') = \|x - x'\|_X$ (i.e. normed spaces).

Furthermore, if $\|\cdot\|_X$ is the l_1 norm, then $D_f \leq C \implies$ in worst case, target dimension is $n^{\Omega(\frac{1}{C^2})}$. That is, there exists a set of n points X, such that for all functions $f : (X, l_1) \to (X', l_1^m)$, with distortion $\leq C$, then m must be at least $n^{\Omega(\frac{1}{C^2})}$ (Brinkman and Charikar 2005).

More recently in 2010, Johnson and Naor (Johnson and Naor 2010) have proposed the following theorem.

Theorem 2.4 *Suppose $(X, \|\cdot\|_x)$ is a complete normed vector space or "Banach Space" such that for any N point subset of X, we can map to $O(\log n)$ dimension subspace of X with $O(1)$ distortion, then every n-dimensional linear subspace of X embeds into l_2 with distortion $\leq 2^{2^{O(\log^* n)}}$.*

Dimension reduction for high dimensional metric data has been an extremely important paradigm in many application areas. In particular, the celebrated Johnson–Lindenstrauss Lemma has played a central role in a plethora of applications. We will now present the Johnson–Lindenstrauss lemma which is a result named after W. B. Johnson and J. Lindenstrauss concerning low-distortion embeddings of points from high-dimensional into low-dimensional Euclidean space. The lemma states that a small set of points in a high-dimensional space can be embedded into a space of much lower dimension in such a way that distances between the points are nearly preserved, the map used for the embedding is at least Lipschitz, and can even be taken to be an orthogonal projection.

2.5.1 Johnson Lindenstrauss Lemma

The Johnson–Lindenstrauss Lemma is a critical tool in the realm of dimensionality reduction and high dimensional approximate computational geometry. It is a classic result which implies that any set of n real vectors can be compressed to $O(\log n)$ dimensions while only distorting pairwise Euclidean distances by a constant factor. It is also employed for data mining in domains that analyse intrinsically high dimensional objects such as images and text. However, while algorithms performing the dimensionality reduction have become increasingly sophisticated, there is little understanding of the behaviour of these embeddings in practice. In many practical instances it is often the case that the high-dimensional data is inherently low dimensional and it is therefore desirable to reduce its dimension close to its inherent dimensionality, which is independent of the size of the data set.

Theorem 2.5 *The Johnson–Lindenstrauss (JL) lemma (Johnson and Lindenstrauss 1984) states that for all $\varepsilon \in (0, \frac{1}{2})$, $\forall x_1, \ldots, x_n \in l_2$, there exists $\Pi \in R^{m \times n}$, $m = O(\frac{1}{\varepsilon^2} \log(n))$ such that*

$for\ all\ i, j, (1 - \varepsilon)\|x_i - x_j\|_2 \le \|\Pi x_i - \Pi x_j\|_2 \le (1 + \varepsilon)\|x_i - x_j\|_2.\ f : (x, l_2) \to$
$(x, l_2^m),\ f(x) = \Pi x$

We also present a distributed Johnson Lindenstrauss theorem

Theorem 2.6 *For all* $0 < \varepsilon, \delta < \frac{1}{2}$, *there exists a distribution* $D_{\varepsilon,\delta}$ *on matrices* $\Pi \in$
$\mathbb{R}^{m \times n}$, $m = O(\frac{1}{\varepsilon^2} \log(\frac{1}{\delta}))$ *such that for all* $x \in \mathbb{R}^n$, *and* Π *drawn from the distribution*
$D_{\varepsilon,\delta}$,

$$\mathbb{P}(\|\Pi x\|_2 \notin [(1 - \varepsilon)\|x\|_2, (1 + \varepsilon)\|x\|_2] < \delta$$

Now we illustrate that the distributional Johnson Lindenstrauss proves Johnson Lindenstrauss.

Claim $DJL \implies JL.$

Proof Set $\delta < \frac{1}{\binom{N}{2}}$ and look at $T = \frac{x_i - x_j}{\|x_i - x_j\|_2}$ for $i < j$. Also see that $|T| = \binom{N}{2}$. Then

$$P(\Pi \text{ doesn't have distortion } (1 + \varepsilon) \text{ on X}) = P(\exists z \in T \text{ such that } \left| \|\Pi z\|_2^2 - 1 \right| \ge \varepsilon)$$

and so by union bound this probability is $\le |T| * \delta < 1$.

Next we collect some notation and basic lemmas we will use (Nelson 2015).

Throughout, for a random variable X, $\|X\|_p$ denotes $(e|X|^p)^{1/p}$. It is known that $\| \cdot \|_p$ is a norm for any $p \ge 1$. In mathematical analysis, Minkowski's inequality establishes that the L^p spaces are normed vector spaces. It is also known $\|X\|_p \le \|X\|_q$ whenever $p \le q$. Whenever we discuss $\| \cdot \|_p$, we will assume $p \ge 1$.

For $F(x) = |x|^p$ $(p \ge 1)$.

In the following part we will use two inequalities. Jensen's Inequality which is an inequality discovered by Danish mathematician Johan Jensen in 1906. Jensen's Inequality appears in many forms depending on the context. The inequality states that the convex transformation of a mean is less than or equal to the mean applied after convex transformation. It is a simple corollary that the opposite is true of concave transformations. The Khintchine inequality, named after Aleksandr Khinchin, is a theorem from probability, and is also frequently used in analysis. The importance of the Rademacher functions and the Khintchine inequality in Functional Analysis lies mainly on the fact of its utility in the study of the geometry of Banach spaces. Additionally, the concern of the Rademacher functions in the theory of functional and trigonometric series and in the theory of Banach spaces is well known and it is commonly attributable to stochastic independence of the Rademacher functions. One of the main manifestations of this stochastic independence is, namely, the Khintchine inequality. Moreover, the Khintchine inequality is an important auxiliary result frequently used to prove results concerning to summability.

Lemma 2.2 *For F convex,* $F(eX) \le eF(X)$.

Lemma 2.3 *For* $1 \le p < q < \infty$, $\|X\|_p \le \|X\|_q$.

Proof Define $f(x) = |x|^{q/p}$. Then f is convex. Thus by Jensen's inequality,

$$(e|X|^p)^{q/p} \leq e|X|^q.$$

Now raise both sides of the inequality to the $1/p$.

Definition 2.2 The Gaussian distribution $\mathcal{N}(0, \sigma^2)$ has density function $f(x) = (2\pi\sigma^2)^{-1/2}e^{-x^2/(2\sigma^2)}$.

Remark 2.2 If $g \sim \mathcal{N}(0, \sigma^2)$ then eg^p for integer p is 0 for p odd and is $\sigma^p(p - 1) \cdot (p - 3) \cdots 1 < (\sigma\sqrt{p})^p$ for p even.

Lemma 2.4 *For any* $p \geq 1$, $x \in \mathbb{R}^n$, *and* (σ_i) *independent Rademachers,*

$$\left\| \sum_i \sigma_i x_i \right\|_p \lesssim \sqrt{p} \cdot \|x\|_2$$

Proof Without loss of generality we can assume p is an even integer. (If not, let q be the smallest even integer larger than p then it suffices to have $\|\sum_i \sigma_i x_i\|_q \lesssim \sqrt{p}\|x\|_2$ since $\|\cdot\|_p \leq \|\cdot\|_q$ by Lemma 2.3.) Consider (g_i) independent Gaussian of mean zero and variance 1. Expand $e(\sum_i \sigma_i x_i)^p$ into a sum of monomials.

$$e\left(\sum_i \sigma_i x_i\right)^p = \sum_{t=1}^{\min\{p,n\}} \sum_{i_1 < i_2 < \cdots < i_t} \sum_{\substack{d_1,\ldots,d_t \geq 1 \\ d_1 + \cdots + d_t = p}} \binom{p}{d_1, \ldots, d_t} \left(\prod_{j=1}^t x_{i_j}^{d_j}\right) \left(e_\sigma \prod_{j=1}^t \sigma_{i_j}^{d_j}\right)$$

$$= \sum_{t=1}^{\min\{p,n\}} \sum_{i_1 < i_2 < \cdots < i_t} \sum_{\substack{d_1,\ldots,d_t \geq 1 \\ d_1 + \cdots + d_t = p}} \binom{p}{d_1, \ldots, d_t} \left(\prod_{j=1}^t x_{i_j}^{d_j}\right) \left(\prod_{j=1}^t e_{\sigma_{i_j}} \sigma_{i_j}^{d_j}\right)$$

Any monomial with odd exponents (i.e. odd d_j) vanishes, as in the Gaussian case. Meanwhile, monomials with all d_j being even have $\prod_{j=1}^t x_{i_j}^{d_j}$ nonnegative and $\prod_{j=1}^t e_{\sigma_{i_j}} \sigma_{i_j}^{d_j} = 1$. Meanwhile if the σ_{i_j} are replaced by Gaussian g_{i_j}, then $\prod_{j=1}^t e_{\sigma_{i_j}} g_{i_j}^{d_j} \geq 1$. Thus the Rademacher pth moment is term-by-term dominated by the Gaussian case and thus $\|\sum_i \sigma_i x_i\|_p \leq \|\sum_i g_i x_i\|_p$. But $\sum_i g_i x_i$ is a Gaussian with mean zero and variance $\|x\|_2^2$.

So far we showed that the Rademacher case is bounded by the Gaussian case. In the following lines you will see another approach. Hereafter we will not require p to be an even integer.

$$\|\sum_i \sigma_i x_i\|_p = \sqrt{\frac{\pi}{2}} \cdot \|\mathbb{e}_g \sum_i \sigma_i |g_i| x_i\|_p \text{ (since } \mathbb{e}|g| = \sqrt{2/\pi})$$

$$\leq \sqrt{\frac{\pi}{2}} \cdot \|\sum_i \sigma_i |g_i| x_i\|_p \text{ (Jensen)}$$

$$= \sqrt{\frac{\pi}{2}} \cdot \|\sum_i g_i x_i\|_p \text{ (}\sigma_i |g_i| \text{ is distributed as } g_i)$$

We further demonstrate a decoupling inequality which will be essential for the proof of Hanson–Wright inequality. Hanson–Wright will be used to prove distributional JL. We use $\| \cdot \|_{L^p(X)}$ to denote $(\mathbb{e}_X | \cdot |^p)^{1/p}$.

Lemma 2.5 *Let $x_1, \ldots, x_n$ be independent and mean zero, and $x_1', \ldots, x_n'$ identically distributed as the x_i and independent of them. Then for any $(a_{i,j})$ and for all $p \geq 1$*

$$\|\sum_{i \neq j} a_{i,j} x_i x_j\|_p \leq 4\|\sum_{i,j} a_{i,j} x_i x_j'\|_p$$

Proof Let $\eta_1, \ldots, \eta_n$ be independent Bernoulli random variables each of expectation $1/2$. Then

$$\|\sum_{i \neq j} a_{i,j} x_i x_j\|_{L^p(x)} = 4 \cdot \|\mathbb{e}_\eta \sum_{i \neq j} a_{i,j} x_i x_j |\eta_i||1 - \eta_j|\|_{L^p(x)}$$

$$\leq 4 \cdot \|\sum_{i \neq j} a_{i,j} x_i x_j \eta_i (1 - \eta_j)\|_{L^p(x,\eta)} \qquad (2.1)$$

Hence there must be some fixed vector $\eta' \in \{0, 1\}^n$ which attains

$$\|\sum_{i \neq j} a_{i,j} x_i x_j \eta_i (1 - \eta_j)\|_{L^p(x,\eta)} \leq \|\sum_{i \in S} \sum_{j \notin S} a_{i,j} x_i x_j\|_{L^p(\eta)}$$

where $S = \{i : \eta_i' = 1\}$. Let x_S denote the $|S|$-dimensional vector corresponding to the x_i for $i \in S$. Then

$$\|\sum_{i \in S} \sum_{j \notin S} a_{i,j} x_i x_j\|_{L^p(x)} = \|\sum_{i \in S} \sum_{j \notin S} a_{i,j} x_i x_j'\|_{L^p(x_S, x_{\bar{S}}')}$$

$$= \|\mathbb{e}_{x_S} \mathbb{e}_{x_{\bar{S}}'} \sum_{i,j} a_{i,j} x_i x_j'\|_{L^p(x_S, x_{\bar{S}}')} \text{ (}\mathbb{e}x_i = \mathbb{e}x_j' = 0)$$

$$\leq \|\sum_{i,j} a_{i,j} x_i x_j'\|_{L^p(x,x')}$$

The Hanson–Wright inequality is equivalent to the statement that there exists a constant $C > 0$ such that for all $\lambda > 0$

$$\mathbb{P}_{\sigma}\left(|\sigma^T A \sigma - \mathrm{e}\sigma^T A \sigma| > \lambda\right) \lesssim e^{-C\lambda^2/\|A\|_F^2} + e^{-C\lambda/\|A\|}. \tag{2.2}$$

Theorem 2.7 ((Hanson and Wright 1971) *For $\sigma_1, \ldots, \sigma_n$ random variables whose distributions are symmetric about zero (independent Rademachers) and $A \in \mathbb{R}^{n \times n}$ real and symmetric, for all $p \geq 1$*

$$\|\sigma^T A \sigma - \mathrm{e}\sigma^T A \sigma\|_p \lesssim \sqrt{p} \cdot \|A\|_F + p \cdot \|A\|.$$

Proof Without loss of generality we suppose that $p \geq 2$ (so that $p/2 \geq 1$). Then

$$\|\sigma^T A \sigma - \mathrm{e}\sigma^T A \sigma\|_p \lesssim \|\sigma^T A \sigma'\|_p \tag{2.3}$$

$$\lesssim \sqrt{p} \cdot \|\|Ax\|_2\|_p \tag{2.4}$$

$$= \sqrt{p} \cdot \|\|Ax\|_2^2\|_{p/2}^{1/2} \tag{2.5}$$

$$\leq \sqrt{p} \cdot \|\|Ax\|_2^2\|_p^{1/2}$$

$$\leq \sqrt{p} \cdot (\|A\|_F^2 + \|\|Ax\|_2^2 - \mathrm{e}\|Ax\|_2^2\|_p)^{1/2} \text{ (triangle inequality)}$$

$$\leq \sqrt{p} \cdot \|A\|_F + \sqrt{p} \cdot \|\|Ax\|_2^2 - \mathrm{e}\|Ax\|_2^2\|_p^{1/2}$$

$$\lesssim \sqrt{p} \cdot \|A\|_F + \sqrt{p} \cdot \|x^T A^T A x'\|_p^{1/2}$$

$$\lesssim \sqrt{p} \cdot \|A\|_F + p^{3/4} \cdot \|\|A^T A x\|_2\|_p^{1/2}$$

$$\lesssim \sqrt{p} \cdot \|A\|_F + p^{3/4} \cdot \|A\|^{1/2} \cdot \|\|Ax\|_2\|_p^{1/2} \tag{2.6}$$

Writing $E = \|\|Ax\|_2\|_p^{1/2}$ and comparing above equations, we see that for some constant $C > 0$,

$$E^2 - Cp^{1/4}\|A\|^{1/2}E - C\|A\|_F \leq 0.$$

Thus E must be smaller than the larger root of the above quadratic equation, implying our desired upper bound on E^2.

The Johnson–Lindenstrauss Lemma is useful because it allows one to project high-dimensional data to a very lower dimensionsal space while approximately preserving all of its metric properties. Several algorithms have runtimes that are exponential in the dimension and with only logarithmic dimension these algorithms become polynomial. Note that the reduced dimension depends only on the number of points and not on the original dimension. The theorem is sometimes called a Distributional Johnson–Lindenstrauss Lemma because of the connection to the J–L Lemma.

Lemma 2.6 *Distributional Johnson-Lindenstrauss Lemma For any integer $n > 1$ and $\varepsilon, \delta \in (0, 1/2)$, there exists a distribution $\mathscr{D}_{\varepsilon,\delta}$ over $\mathbb{R}^{m \times n}$ for $m \lesssim \varepsilon^{-2} \log(1/\delta))$ such that for any $x \in \mathbb{R}^n$ of unit Euclidean norm,*

$$\mathop{\mathbb{P}}_{\Pi \sim \mathscr{D}_{\varepsilon,\delta}} (|\,\|\Pi x\|_2^2 - 1| > \varepsilon) < \delta$$

Proof Write $\Pi_{i,j} = \sigma_{i,j}/\sqrt{m}$, where the $\sigma_{i,j}$ are independent Rademachers. Also overload σ to mean these Rademachers arranged as a vector of length mn, by concatenating rows of Π. See

$$\Pi x = A_x \sigma, \text{ implying } \|\Pi x\|_2^2 = \|A_x \sigma\|_2^2$$

where

$$A_x = \frac{1}{\sqrt{m}} \cdot \begin{bmatrix} -x^T- & 0 & \cdots & 0 \\ 0 & -x^T- & \cdots & 0 \\ \vdots & \vdots & & \vdots \\ 0 & 0 & \cdots & -x^T- \end{bmatrix}. \tag{2.7}$$

Thus

$$\mathbb{P}(|\,\|\Pi x\|_2^2 - 1| > \varepsilon) = \mathbb{P}(|\,\|A_x\sigma\|_2^2 - \mathrm{e}\|A_x\sigma\|_2^2| > \varepsilon),$$

where the right-hand side is taken by the Hanson–Wright inequality with $A = A_x^T A_x$ (from (2.2)). Note that A is a block-diagonal matrix with each block equaling $(1/m)xx^T$, and thus $\|A\| = \|x\|_2^2/m = 1/m$. We also have $\|A\|_F^2 = 1/m$. Thus Hanson–Wright allows

$$\mathbb{P}(|\,\|\Pi x\|_2^2 - 1| > \varepsilon) \lesssim e^{-C\varepsilon^2 m} + e^{-C\varepsilon m},$$

which for $\varepsilon < 1$ is at most δ for $m \gtrsim \varepsilon^{-2} \log(1/\delta)$.

2.5.2   *Lower Bounds on Dimensionality Reduction*

Lower bound on dimentionality reduction was initially proved by Jayram and Woodruff (Jayram and Woodruff 2013). Thereafter Kane et al. (Kane et al. 2011) also showed lower bound. The lower bound states that any (ε, δ)-DJL for $\varepsilon, \delta \in (0, 1/2)$ must have
$m = \Omega(min\{n, \varepsilon^{-2} \log(\frac{1}{\delta})\})$.

Since for all x we have the probabilistic guarantee $\mathbb{P}_{\Pi \sim \mathbb{D}_{\varepsilon,\delta}}[|\,\|\Pi x\|_2^2 - 1| < max\{\varepsilon, \varepsilon^2\}] < \delta$, then it is true also for any distribution over x. If we select x according to the uniform distribution over the sphere. Then this implies that there exists a matrix Π such that $\mathbb{P}_x[|\,\|\Pi x\|_2^2 - 1| < \varepsilon] < \delta$. It is proved that this cannot happen for any fixed matrix $\Pi \in \mathbb{R}^{m \times n}$ unless m satisfies the lower bound.

In case of linear maps, a lower bound of Kasper and Nelson (Larsen and Nelson 2014) shows us that m should be at least $\Omega(min\{n, \varepsilon^{-2} \log |X|\})$. The hard set is shown to exist via the probabilistic method, and is constructed by taking the union of $\{0, e_1, \ldots, e_n\}$ in $\mathbb{R}^n$ together with plus sufficiently many random vectors. Now we assume a fine finite net that approximates all possible linear maps that work for the simplex (i.e. $\{e_1, \ldots, e_n\}$) and then argue that for each linear map in the net then with some good probability there exists a point from the random ones such that its length is not preserved by the map. Therefore, approximating the set of linear maps and tuning the parameters, we can take a union bound over all matrices in the net.

This lower bound has been proved by Alon (Alon 2003). It shows that m must be at least $\Omega(min\{n, \varepsilon^{-2} \frac{\log n}{\log(\frac{1}{\varepsilon})}$ to preserve distances between the set of points $X = \{0, e_1, \ldots, e_n\}$. The hard set is the simplex $X = \{0, e_1, \ldots, e_n\}$. Let f be the mapping. Transform the embedding so that $f(0) = 0$ (i.e. translate transform the embedding so that each point $x \in X$ is actually mapped to $f(x) - f(0)$; this does not affect any distances). Now write $f(e_i) = v_i$. Then we have that $\|v_i\| = 1 \pm \varepsilon$ since f preserves the distance from e_i to 0. We also have for $i \neq j$ that $\|v_i - v_j\| = \sqrt{2}(1 \pm \varepsilon)$. This implies since $\|v_i - v_j\| = \|v_i\|^2 + \|v_j\|^2 - 2\langle v_i, v_j \rangle$ we get that $\langle v_i, v_j \rangle = O(\varepsilon)$. Setting $w_i = \frac{v_i}{\|v_i\|}$ we have that $\|w_i\| = 1$ and $\langle w_i, w_j \rangle = O(\varepsilon)$. Let Π be the matrix that has w_i as columns. Then observe that $A = \Pi^T \Pi$ is a matrix with 1 on the diagonal and elements with absolute value at most ε everywhere. We have that $rank(A) = rank(\Pi^T \Pi) = rank(\Pi) \leq m$. Let us consider the following lemma that unfolds the problem for $\varepsilon = \frac{1}{\sqrt{n}}$ and further bootstrap it to work for all values of ε.

Lemma 2.7 *A real symmetric matrix that is $\frac{1}{\sqrt{n}}$-near to the identity matrix, i.e. its diagonals are 1 and off-diagonals are in $[-1/\sqrt{n}, 1/\sqrt{n}]$, has $rank(A) \geq \Omega(n)$.*

Proof Let $\lambda_1, \ldots, \lambda_r$ be the non-zero eigenvalues of A, where $r = rank(A)$. By Cauchy-Schwarz inequality, we have $r \geq \frac{(\sum_{i=1}^r \lambda_i)^2}{\sum_{i=1}^r \lambda_i^2}$. By using linear algebra the numerator is the trace of A squared and the denominator is the Frobenius norm of A squared. We have that $tr(A) = n$ and $\|A\|_F^2 \leq n + n(n-1)\varepsilon^2$. Pushing it into the inequality along with the fact that $\varepsilon = \frac{1}{\sqrt{n}}$ we prove the lemma.

Theorem 2.8 *A real symmetric matrix that is ε-near to the identity matrix must have $rank(A) \leq min\{n, \varepsilon^{-2} \frac{\log n}{\log \frac{1}{\varepsilon}}\}$.*

Proof Define the matrix $A^{(k)}$ such that $(A^{(k)})_{ij} = a_{ij}^k$. We will build our proof on the following claim: It holds that $rank(A^{(k)} \leq \binom{r+k-1}{k}$ where $r = rank(A)$. Assuming that the claim is true we pick k to be the closest integer to $\log n_{\varepsilon^{-1}} \sqrt{n}$. Thus $\varepsilon^k \leq \frac{1}{\sqrt{n}}$, so we have that $\Omega(n) \leq r(A^{(k)}) \leq \binom{r+k-1}{k}$. Using the fact that $\binom{n}{k} \leq (enk^{-1})^k$ and walking thought the calculations we can get the desired result.

What remains is to prove the claim. Let $t_1, \ldots, t_r$ be the row-space of A. This means that $\forall i \exists \beta \in \mathbb{R}^r$ such that $a_i = \sum_{q=1}^r \beta_q t_q$. Then observe that $(A^{(k)})_{ij} = a_{ij}^k = (\sum_{q=1}^r \beta e_q (t_q)_j)^k = \sum_{q_1, \ldots, q_k} \Pi_{z=1}^k \beta_{q_z} \Pi_{z=1}^k t_{q_z}$. It is easy to see that each vector of this form is a linear combination of vectors of the form

$(\Pi_{z=1}^{y}(t_{q_z})_1^{d_z^1}, \Pi_{z=1}^{y}(t_{q_z})_2^{d_z^2}, \ldots)$. where $\sum d_z^i = k$. This is a standard combinatorics problem of putting r balls into k bins with repetition, so the answer is $\binom{r+k-1}{k}$.

Given a subset T of the unit sphere- for example $\{T = \frac{x-y}{\|x-y\|}, x, y \in X\}$- ideally we would like that $\forall x \in T, |\|\Pi x\|^2 - 1| < \varepsilon$. We want that $\mathbb{P}_\Pi(sup_{x \in T}|\|\Pi x\|^2 - 1| > \varepsilon) < \delta$.

Definition 2.3 The Gaussian mean width of a set T is defined as $g(T) = \mathbb{E}_g sup_{x \in T} \langle g, x \rangle$.

Suppose that $\Pi_{ij} = \frac{\pm 1}{\sqrt{m}}$ for random signs, with $m \geq \Omega(\varepsilon^{-2}(g^2(T) + \log\frac{1}{\delta}))$. Then we have that $\mathbb{P}_\Pi(sup_{x \in T}|\|\Pi x\|^2 - 1| > \varepsilon) < \delta$. Actually, we just need a distribution that decays as fast as a Gaussian, has variance one and zero mean.

Let us give a simple example of the Gaussian mean width. For example, if T is the simplex then we have that $g(T) = \|g\|_\infty$ which is roughly equal to $\log n$ by standard computations. Actually, what Gordon's theorem tells us is that if the vectors of T have a nice geometry then one can improve upon Johnson–Lindenstrauss. The more well-clustered the vectors are, the lower dimension one can achieve.

We continue with the following claim: $\forall T$ which is a subset of the unit sphere, $g(T) \leq O(\sqrt{\log N})$, where $N = |T|$.

Proof Define $Q_i = |\langle g_i, x_i \rangle|$, where $T = \{x_1, \ldots, x_N\}$. Then

$$g(T) \leq \mathbb{E}_g max\{Q_1, \ldots, Q_N\} = \int_0^\infty \mathbb{P}_g(max\{Q_1, \ldots, Q_N\} > u)du =$$

$$= \int_0^{2\sqrt{\log n}} \mathbb{P}_g(max\{Q_1, \ldots, Q_N\} > u)du + \int_0^{2\sqrt{\log n}} \mathbb{P}_g(max\{Q_1, \ldots, Q_N\} > u)du \leq$$

$$\leq 2\sqrt{\log n} + \int_{2\sqrt{\log n}}^\infty \mathbb{P}_g(\exists Q_i \geq u)du \leq 2\sqrt{\log n} + \int_{2\sqrt{\log n}}^\infty \sum_i \mathbb{P}_g(Q_i \geq u)du \leq$$

$$2\sqrt{\log N} + \int_{2\sqrt{\log n}}^\infty N e^{-u^2/2} \leq 2\sqrt{\log n} + O(1)$$

$g(T) \leq \sqrt{\log |T|}$, as we have illustrated. In fact if every vector in T has norm at most α, then one gets $g(T) \lesssim \alpha\sqrt{\log |T|}$. Let $T' \subset T$ be such that $\forall x \in T, \exists x' \in T'$ such that $\|x - x'\| \leq \varepsilon$. That is, T' is an ε-net of T. This implies that $\langle g, x \rangle = \langle g, x' \rangle + \langle g, x - x' \rangle \leq \|g\|_2 \varepsilon$, which implies that $g(T) \leq g(T') + \varepsilon\mathbb{E}\|g\|_2^2 \leq g(T') + e(\mathbb{E}\|g\|_2^2)^{\frac{1}{2}} \leq g(T') + \varepsilon\sqrt{n} \leq O(\sqrt{\log |T'|}) + \varepsilon\sqrt{n}$. Thus if T is covered well by a small net, one can get a better bound. Let $T \subset T_1 \subset T_2 \subset \cdots \subset T$, such that T_r is a (2^{-r})-net of T (we are assuming every vector in T has at most unit norm). Then $x = x^{(0)} + (x^{(1)} - x^{(0)}) + (x^{(2)} - x^{(1)}) + \cdots$. Then we have

$$e\sup_{x\in T}\langle g, x\rangle \leq e\sup_{x\in T}\langle g, x^{(0)}\rangle + \sum_{r=1}^{\infty} e\sup_{x\in T}\langle g, x^{(r)} - x^{(r-1)}\rangle$$

$$\lesssim \log^{1/2}|T_0| + \sum_{r=1}^{\infty}(\sup_{x\in T}\|x^{(r)} - x^{(r-1)}\|) \cdot \log^{1/2}(|T_r| \cdot |T_{r-1}|)$$

$$\lesssim \log^{1/2}|T_0| + \sum_{r=1}^{\infty}\frac{1}{2^r} \cdot \log^{1/2}|T_r|.$$

The last inequality holds since by the triangle inequality, $\|x^{(r)} - x^{(r-1)}\| \leq \|x^{(r)} - x\| + \|x^{(r-1)} - x\| \leq 3/2^{r-1}$. Furthermore, $|T_{r-1}| \leq |T_r|$, so $\log(|T_r| \cdot |T_{r-1}|) \leq \log(|T_r|^2) = 2\log|T_r|$.

Thus $g(T) \leq \sum_{r=0}^{\infty} 2^{-r}\log^{\frac{1}{2}}|T_r| = \sum_{r=0}^{\infty} 2^{-r}N(T_2, \|\dot{\|}_2, 2^{-r})$, where $N(T, d, \varepsilon)$ is the size of the best ε-net of T under metric d. Bounding this sum by an integral, we have that $g(T)$ is at most a constant factor times $\int_0^{\infty} \log^{\frac{1}{2}} N(T, \|\cdot\|_2, u)du$. This inequality is called Dudley's inequality.

Write $S_0 \subset S_1 \subset \cdots \subset T$, such that $|S_0| = 1$ and $\|S_s\| \leq 2^{2^s}$. One can show that the Dudley bound is in fact

$$\inf_{\{S_s\}_{s=0}^{\infty}} \sum_{s=0}^{\infty} 2^{s/2}\sup_{x\in T} d_{\|\cdot\|_2}(x, S_s).$$

Write

$$\gamma_2(T) = \inf_{\{S_s\}_{s=0}^{\infty}} \sup_{x\in T} \sum_{s=0}^{\infty} 2^{s/2}d_{\|\cdot\|_2}(x, S_s).$$

It was shown by Fernique (Fernique 1975) that $g(T) \lesssim \gamma_2(T)$ for all T. Talagrand proved that in (Talagrand 1996) the lower bound is also true, and hence $g(T) = \Theta(\gamma_2(T))$; this is known as the "majorizing measures" theorem.

2.5.3 *Dimensionality Reduction for k-Means Clustering*

Clustering is ubiquitous in science and engineering with various application areas ranging from social science and medicine to the biology and the web. The most well-known clustering algorithm is the so-called k-means algorithm (Lloy 1982). This method is an iterative expectation-maximization type approach that attempts to address the following objective. Given a set of Euclidean points and a positive integer k corresponding to the number of clusters, split the points into k clusters so that the total sum of the squared Euclidean distances of each point to its nearest cluster center is minimized.

In recent years, the high dimensionality of enormous datasets has created a significant challenge to the design of efficient algorithmic solutions for k-means clustering.
First, ultra-high dimensional data force existing algorithms for k-means clustering
to be computationally inefficient, and second, the existence of many irrelevant features may not allow the identification of the relevant underlying structure in the
data (Guyon et al. 2005). Researchers have addressed these obstacles by introducing feature selection and feature extraction techniques. Feature selection selects a
(small) subset of the actual features of the data, whereas feature extraction constructs
a (small) set of artificial features based on the original features.

Consider m points $P = \{\mathbf{p}_1, \mathbf{p}_2, \ldots, \mathbf{p}_m\} \subseteq \mathbb{R}^n$ and an integer k denoting the
number of clusters. The objective of k-means is to find a k-partition of P such that
points that are "close" to each other belong to the same cluster and points that are
"far" from each other belong to different clusters. A k-partition of P is a collection
$S = \{S_1, S_2, \ldots, S_k\}$ of k non-empty pairwise disjoint sets which covers P. Let
$s_j = |S_j|$ be the size of $S_j (j = 1, 2, \ldots, k)$. For each set S_j, let $\mu_j \in \mathbb{R}_n$ be its
centroid:

$$\mu_j = \frac{\sum_{p^i \in S_j} \mathbf{p}_i}{s_j}. \tag{2.8}$$

The k-means objective function is

$$F(P, S) = \sum_{i=1}^{m} \|\mathbf{p}_i - \mu(\mathbf{p}_i)\|_2^2, \tag{2.9}$$

where $\mu(\mathbf{p}_i) \in \mathbb{R}_n$ is the centroid of the cluster to which $\mathbf{p}_i$ belongs. The objective
of k-means clustering is to compute the optimal k-partition of the points in $\mathscr{P}$,

$$S_{opt} = \arg \min_S F(P, S). \tag{2.10}$$

Now, the goal of dimensionality reduction for k-means clustering is to construct
points

$$\hat{P} = \{\hat{\mathbf{p}}_1, \hat{\mathbf{p}}_2, \ldots, \hat{\mathbf{p}}_m\} \subseteq \mathbb{R}^r \tag{2.11}$$

(for some parameter $r \ll n$)

Thus $\hat{P}$ approximates the clustering structure of P. Dimensionality reduction via
feature selection constructs the $\hat{\mathbf{p}}_i$'s by selecting actual features of the corresponding
$\mathbf{p}_i$'s, whereas dimensionality reduction via feature extraction constructs new artificial
features based on the original features.

Assume that the optimum k-means partition of the points in $\hat{P}$ has been computed.

$$\hat{S}_{opt} = \arg \min_S F(\hat{P}, S). \tag{2.12}$$

A dimensionality reduction algorithm for k-means clustering constructs a new set $\hat{P}$ such that

$$F(P, \hat{S}_{opt}) \leq \gamma \cdot F(P, S_{opt}) \tag{2.13}$$

where $\gamma > 0$ is the approximation ratio of $\hat{S}_{opt}$. Then again, we need that computing an optimal partition $\hat{S}_{opt}$ on the projected low-dimensional data and plugging it back to cluster the high dimensional data, will imply a γ factor approximation to the optimal clustering.

2.6 Gordon's Theorem

Given a set S, what is the measure of complexity of S that explains how many dimensions one needs to take on the projection to approximately preserve the norms of points in S. This is captured by Gordon's theorem. Oymak, Recht and Soltanolkotabi (Oymak et al. 2015) have showed that with right parameters the Distributional Johnson Lindenstrauss (DJL) lemma implies Gordon's theorem. Basically take a DJL $\varepsilon' = \frac{\varepsilon}{\gamma_2^2(T)}$ then for $m \geq \varepsilon^{-2}(\gamma_2^2(T) \log \frac{1}{\delta})$(where we hide constants in the inequalities) we take the guarantee for Gordon's Theorem. The proof works by preserving the sets S_s (plus differences and sums of vectors in these sets) at different scales. The result is not exactly optimal because it is known $m = O(\varepsilon^{-2}(\gamma_2^2(T) + \log(1/\delta)))$ suffices (see for example (Dirksen 2015; Mendelson et al. 2007), but it provides a nice reduction from Gordon's theorem to DJL. A classical result due to Gordon characterizes the precise trade-off between distortion, "size" of the set and the amount of reduction in dimension for a subset of the unit sphere.

The main result is summarized in the following theorem.

Theorem 2.9 (Oymak et al. 2015) *Define* $L = \lceil \log n \rceil$, $\tilde{\varepsilon} = \varepsilon/(c\gamma_2(T))$, $\tilde{\varepsilon}_r = \max\{2^{r/2}\tilde{\varepsilon}, 2^{r/2}\tilde{\varepsilon}^2\}$, $\delta_r = \frac{\delta}{C2^r 8^{2^r}}$. *Let* $T \subset S^{n-1}$. *Then if* D *satisfies* $(\varepsilon_r, \delta_r)$-DJL *for* $r = 0, \ldots, L$, *then*

$$P_{\Pi \sim D}\left(\sum_{x \in T} \left| \|\Pi x\|_2^2 - 1 \right| > \varepsilon \right) < \delta$$

To illustrate why this implies Gordon's theorem, we take the random sign matrix, e.g., $\Pi_{ij} = \frac{\sigma_{ij}}{\sqrt{m}}$. We know that this matrix satisfies $(\tilde{\varepsilon}, \tilde{\delta})$-DJL for $m \gtrsim \frac{\log(1/\tilde{\delta})}{\tilde{\varepsilon}^2}$, which equals $\frac{2^r \log(1/\tilde{\delta})}{(2^{r/2}\tilde{\varepsilon})^2} \geq \frac{\log(1/\delta_r)}{\varepsilon_r^2}$ for all r. The theorem therefore applies and so we see that we get an (ε, δ) guarantee with $m \gtrsim \log(1/\delta)/\tilde{\varepsilon}^2 \approx \frac{\gamma_2^2(T)}{\varepsilon^2} \log(1/\delta)$. And since $\gamma_2(T) \simeq g(T)$, this is approximately $\frac{g^2(T) \log(1/\delta)}{\varepsilon^2}$, which gives Gordon's theorem. Different proofs yield that $\frac{g^2(T) + \log(1/\delta)}{\varepsilon^2}$ obviously suffices.

To prove the above theorem, the lemma below suffices.

Lemma 2.8 *For a given set T, let T_r be the sequence that achieves the infimum in the definition of γ_2. To achieve $\sup_{x \in T} \left| \|\Pi x\|_2^2 - 1 \right| < \varepsilon$, it suffices that for all $r = 0, \ldots, L$, the following hold simultaneously for all $r \in [L]$.*
For all $v \in T_{r-1} \cup T_r \cup (T_{r-1} - T_r)$,

$$\|\Pi v\| \le (1 + 2^{r/2} \tilde{\varepsilon}) \|v\| \tag{2.14}$$

For all $v \in T_{r-1} \cup T_r \cup (T_{r-1} - T_r)$,

$$\left| \|\Pi v\|^2 - \|v\|^2 \right| \le \max\{2^{r/2} \tilde{\varepsilon}, 2^r \tilde{\varepsilon}^2\} \cdot \|v\|^2 \tag{2.15}$$

For all $u \in T_{r-1}$ and $v \in T_r - \{u\}$,

$$\left| \langle \Pi u, \Pi v \rangle - \langle u, v \rangle \right| \le \max\{2^{r/2} \tilde{\varepsilon}, 2^r \tilde{\varepsilon}^2\} \cdot \|u\| \cdot \|v\| \tag{2.16}$$

We also have
$$\|\Pi\| \le 1 + (1/4) 2^{L/2} \tilde{\varepsilon} \tag{2.17}$$

Obviously, the first three conditions hold with high probability since they are all JL-type conditions. The third one is a bit less obvious since it is about dot products instead of norms. But notice that $\|u + v\|^2 - \|u - v\|^2 = 4\langle u, v \rangle$. So if $\|u\| = \|v\| = 1$, then Π preserving $u + v$ and $u - v$ means that $\langle \Pi u, \Pi v \rangle = \frac{1}{4}(\|\Pi u + \Pi v\|^2 - \|\Pi u + \Pi v\|^2) = \langle u, v \rangle \pm O(\varepsilon)$. If u and v don't have unit norm you can scale them to achieve the above condition. So the third condition also follows from the DJL premise.

We now argue that the lemma suffices to prove the theorem.

Claim *Lemma 2.8 implies Theorem 2.9.*

Proof Define $\tilde{L} = \lceil \log(1/\tilde{\varepsilon}^2) \rceil \le L$ Fix $x \in T$. We will show

$$\left| \|\Pi x\|^2 - \|x\|^2 \right| < \varepsilon$$

Define $e_r(T) = d(x, T_r)$, and define

$$\tilde{\gamma}_2(T) = \sum_{r=1}^{L} 2^{r/2} \cdot e_r(T).$$

So, $\tilde{\gamma}_2(T) \le \gamma_2(T)$.
    Now define

$$z_r = \operatorname{argmin}_{y \in T_r} \|x - y\|_2$$

$$\left| \|\Pi x\|^2 - \|x\|^2 \right| \le \left| \|\Pi z_{\tilde{L}}\|^2 - \|z_{\tilde{L}}\|^2 \right| + \left| \|\Pi x\|^2 - \|\Pi z_{\tilde{L}}\|^2 \right| + \left| \|x\|^2 - \|z_{\tilde{L}}\|^2 \right|$$

$$\le \underbrace{\left| \|\Pi z_0\|^2 - \|z_0\|^2 \right|}_{\alpha} + \underbrace{\left| \|\Pi x\|^2 - \|\Pi z_{\tilde{L}}\|^2 \right|}_{\beta} + \underbrace{\left| \|x\|^2 - \|z_{\tilde{L}}\|^2 \right|}_{\Gamma}$$

$$+ \underbrace{\sum_{r=1}^{\tilde{L}} \left(\left| \|\Pi z_r\|^2 - \|z_r\|^2 \right| - \left| \|\Pi z_{r-1}\|^2 - \|z_{r-1}\|^2 \right| \right)}_{\Delta} \quad (2.18)$$

We now proceed by bounding each of the four terms as follows.

Case α:

We have $\alpha \le \max\{\tilde{\varepsilon}, \tilde{\varepsilon}^2\} \le \tilde{\varepsilon}$.

Case β:

We have

$$\left| \|\Pi x\|^2 - \|\Pi z_{\tilde{L}}\|^2 \right| = \left| \|\Pi x\| - \|\Pi z_{\tilde{L}}\| \right| \cdot \left| \|\Pi x\| + \|\Pi z_{\tilde{L}}\| \right|$$

$$\le \left| \|\Pi x\| - \|\Pi z_{\tilde{L}}\| \right| \cdot \left(\left| \|\Pi x\| - \|\Pi z_{\tilde{L}}\| \right| + 2 \cdot \|\Pi z_{\tilde{L}}\| \right)$$

$$= \left| \|\Pi x\| - \|\Pi z_{\tilde{L}}\| \right|^2 + 2 \left| \|\Pi x\| - \|\Pi z_{\tilde{L}}\| \right| \cdot \|\Pi z_{\tilde{L}}\| \quad (2.19)$$

We thus need to bound $\left| \|\Pi x\| - \|\Pi z_{\tilde{L}}\| \right|$ and $\|\Pi z_{\tilde{L}}\|$.

We have $\|\Pi z_{\tilde{L}}\| \le (1 + 2^{\tilde{L}/2}\tilde{\varepsilon}) \le 2$.

Next, we consider

$$\left| \|\Pi x\| - \|\Pi z_{\tilde{L}}\| \right| = \left| \|\Pi x\| - \|\Pi z_L\| + \|\Pi z_L\| - \|\Pi z_{\tilde{L}}\| \right|$$

$$\le \|\Pi(x - z_L)\| + \|\Pi(z_L - z_{\tilde{L}})\|$$

$$\le \|\Pi\| \cdot \|x - z_L\| + \left\| \sum_{r=\tilde{L}+1}^{L} \Pi(z_r - z_{r-1}) \right\|$$

$$\le \|\Pi\| \cdot e_L(T) + \sum_{r=\tilde{L}+1}^{L} \|\Pi(z_r - z_{r-1})\| \quad (2.20)$$

Now $\|\Pi\| \le \frac{1}{4}2^{L/2}\tilde{\varepsilon} + 1$. $\|\Pi(z_r - z_{r-1})\| \le (1 + 2^{r/2}\tilde{\varepsilon})\|z_r - z_{r-1}\|$. Thus, using $2^{r/2}\tilde{\varepsilon} \ge 1$ for $r > \tilde{L}$,

$$\le (\frac{1}{4}2^{L/2}\tilde{\varepsilon} + 1)e_L(T) + \sum_{r=\tilde{L}+1}^{L} (1 + 2^{r/2}\tilde{\varepsilon}\|z_r - z_{r-1}\|$$

$$\le (\frac{1}{4}2^{L/2}\tilde{\varepsilon} + 1)e_L(T) + \sum_{r=\tilde{L}+1}^{L} (1 + 2^{r/2}\tilde{\varepsilon})\|z_r - z_{r-1}\|$$

$$\leq \frac{5}{4} 2^{L/2} \tilde{\varepsilon} e_L(T) + \sum_{r=\tilde{L}+1}^{L} 2^{r/2+1} \tilde{\varepsilon} \|z_r - z_{r-1}\|$$

$$\leq \frac{5}{4} 2^{L/2} \tilde{\varepsilon} e_L(T) + 4\sqrt{2} \tilde{\varepsilon} \sum_{r=\tilde{L}+1}^{L} 2^{(r-1)/2} \cdot e_{r-1}(T)$$

$$\leq 4\sqrt{2} \tilde{\varepsilon} \sum_{r=\tilde{L}}^{L} 2^{r/2} \cdot e_r(T)$$

$$\leq 4\sqrt{2} \tilde{\varepsilon} \cdot \tilde{\gamma}_2(T) \tag{2.21}$$

In order to conclude, we write

$$\beta \leq 32 \tilde{\varepsilon}^2 \tilde{\gamma}_2^2(T) + 16\sqrt{2} \tilde{\varepsilon} \tilde{\gamma}_2(T)$$

Case Γ:
$2^{r/2} \tilde{\varepsilon} \geq 1/\sqrt{2}$ for $r \geq \tilde{L}$. Thus

$$\big| \|x\| - \|z_{\tilde{L}}\| \big| \leq e_{\tilde{L}}(T)$$
$$\leq \sqrt{2} \cdot 2^{\tilde{L}/2} \tilde{\varepsilon} e_{\tilde{L}}(T)$$
$$\leq \sqrt{2} \tilde{\varepsilon} \cdot \tilde{\gamma}_2(T).$$

Thus

$$\Gamma = \big| \|x\|^2 - \|z_{\tilde{L}}\|^2 \big|$$
$$= \big| \|x\| - \|z_{\tilde{L}}\| \big| \cdot \big| \|x\| + \|z_{\tilde{L}}\| \big|$$
$$\leq \big| \|x\| - \|z_{\tilde{L}}\| \big|^2 + 2 \big| \|x\| - \|z_{\tilde{L}}\| \big| \cdot \|z_{\tilde{L}}\|$$
$$\leq 2\tilde{\varepsilon}^2 \cdot \tilde{\gamma}_2^2(T) + 2\sqrt{2} \tilde{\varepsilon} \cdot \tilde{\gamma}_2(T)$$

Case Δ:
By the triangle inequality, for any $r \geq 1$

$$\big| \|\Pi z_r\|^2 - \|z_r\|^2 \big| = \big| \|\Pi(z_r - z_{r-1}) + \Pi z_{r-1}\|^2 - \|(z_r - z_{r-1}) + z_{r-1}\|^2 \big|$$
$$= \big| \|\Pi(z_r - z_{r-1})\|^2 + \|\Pi z_{r-1}\|^2 + 2\langle \Pi(z_r - z_{r-1})), \Pi z_{r-1} \rangle \tag{2.22}$$
$$- \|z_r - z_{r-1}\|^2 - \|z_{r-1}\|^2 - 2\langle z_r - z_{r-1}, z_{r-1} \rangle \big|$$
$$\leq \big| \|\Pi(z_r - z_{r-1})\|^2 - \|z_r - z_{r-1}\|^2 \big| + \big| \|\Pi z_{r-1}\|^2 - \|z_{r-1}\|^2 \big|$$
$$+ 2 \big| \langle \Pi(z_r - z_{r-1})), \Pi z_{r-1} \rangle - \langle z_r - z_{r-1}, z_{r-1} \rangle \big|. \tag{2.23}$$

We have

$$\big| \|\Pi(z_r - z_{r-1})\|^2 - \|z_r - z_{r-1}\|^2 \big| \leq \max\{2^{r/2} \tilde{\varepsilon}, 2^r \tilde{\varepsilon}^2\} \cdot 2e_{r-1}^2(T) \leq 2^{r/2+2} \tilde{\varepsilon} e_{r-1}^2(T),$$

with the second inequality holding since $2^{r/2}\tilde{\varepsilon} \le 1$ for $\le \tilde{L}$.

So, we also have

$$|\langle \Pi(z_r - z_{r-1}), \Pi z_{r-1} \rangle - \langle z_r - z_{r-1}, z_{r-1} \rangle| \le 2^{r/2+1}\tilde{\varepsilon}e_{r-1}.$$

Therefore

$$|\|\Pi z_r\|^2 - \|z_r\|^2| - |\|\Pi z_{r-1}\|^2 - \|z_{r-1}\|^2| \le \tilde{\varepsilon}(2e_{r-1}(T) + 4e_{r-1}^2(T))2^{r/2}$$

Considering $e_r(T) \le 1$ for all r,

$$\Delta \le 10\tilde{\varepsilon} \left(\sum_{r=1}^{\tilde{L}} 2^{r/2}e_{r-1}(T) \right)$$

$$= 10\sqrt{2}\tilde{\varepsilon} \left(\sum_{r=0}^{\tilde{L}-1} 2^{r/2}e_r(T) \right)$$

$$\le 10\sqrt{2}\tilde{\varepsilon}\tilde{\gamma}_2(T).$$

Finally we arrive at

$$|\|\Pi x\|^2 - \|x\|^2| \le \tilde{\varepsilon} + 32\tilde{\varepsilon}^2\tilde{\gamma}_2^2(T) + 16\sqrt{2}\tilde{\varepsilon}\tilde{\gamma}_2(T) + 8\tilde{\varepsilon}^2\tilde{\gamma}_2^2(T) + 2\sqrt{2}\tilde{\varepsilon}\tilde{\gamma}_2(T) + 10\sqrt{2}\tilde{\varepsilon}\tilde{\gamma}_2(T)$$
$$= \tilde{\varepsilon} + 28\sqrt{2}\tilde{\varepsilon}\tilde{\gamma}_2(T) + 40\tilde{\varepsilon}^2\tilde{\gamma}_2^2,$$

where ε for $\tilde{\varepsilon} \le \varepsilon/(84\sqrt{2}\tilde{\gamma}_2(T))$.

2.7 Johnson–Lindenstrauss Transform

Enormous amount of data stored and manipulated on computers can be represented as points in a high-dimensional space. However, the algorithms for working with such data tend to become bogged down very rapidly as dimension increases. It is therefore desirable to reduce the dimensionality of the data in a way that preserves its relevant structure. The Johnson Lindenstrauss lemma is an important result in this respect.

Fast Johnson–Lindenstrauss Transform (FJLT) was introduced by Ailon and Chazelle in 2006 (Ailon and Chazelle 2009). We will discuss that transform in the next section. Another approach is to build a distribution supported over matrices that are sparse.

In high-dimensional computational geometry problem, one can employ Johnson Lindenstrauss transform to speed up the algorithm in two steps: (1) apply a Johnson–Lindenstrauss (JL) map Π to reduce the problem to low dimension m, then (2) solve the lower-dimensional problem.

As m is made smaller, (2) becomes faster. Yet, one would also use step (1) to be as fast as possible. The dimensionality reduction has been a dense matrix-vector multiplication (Nelson 2015).

There are two possible ways of doing this: one is to make Π sparse. We saw in pset 1 that this sometimes works: we replaced the AMS sketch with a matrix each of whose columns has exactly 1 non-zero entry. The other way is to make Π structured, i.e., it's still dense but has some structure that allows us to multiply faster.

One way to speed up JL is to make Π sparse. If Π has s non-zero entries per column, then Πx can be multiplied in time $O(s \cdot \|x\|_0)$, where $\|x\|_0 = |\{i : x_i \neq 0\}|$. The aim is then to make s, m as small as possible.

From (Achlioptas 2003)

$$\Pi_{ij} = \begin{cases} +/\sqrt{m/3} & \text{w.p. } \frac{1}{6} \\ -\sqrt{m/3} & \text{w.p. } \frac{1}{6} \\ 0 & \text{w.p. } \frac{2}{3} \end{cases}$$

and gives DJL, with constant factors. But it provides a factor-3 speed-up since in expectation only one third of the entries in Π are non-zero. On the other hand, (Matousek 2008) proved that if Π has independent entries then you can't speed things up by more than a constant factor.

The first to exhibit a Π without independent entries and therefore to break this lower bound were (Dasgupta et al. 2010), who got $m = O(\frac{1}{\varepsilon^2} \log(1/\delta))$, $s = \tilde{O}(\frac{1}{\varepsilon} \log^3(1/\delta))$ nonzeros per column of Π. So depending on the parameters this could either be an improvement or not.

Now let us see (Kane and Nelson 2014) that you can take $m = O(\frac{1}{\varepsilon^2} \log(1/\delta))$ and $s = O(\frac{1}{\varepsilon} \log(1/\delta))$, a strict improvement by choosing exactly s entries in each column of Π to have non-zero entries and then choosing the signs of those entries at random and normalizing appropriately. Instead you can separate each column of Π up into s blocks of size m/s, and set exactly 1 non-zero entry in each block. The resulting matrix is exactly the count sketch matrix.

The analysis employs Hanson–Wright inequality. For dense Π, we have seen that $\Pi x = A_x \sigma$ where A_x was an $m \times mn$ matrix whose ith row had $x^T/\sqrt{m}$ in the ith block of size n and zeros elsewhere. Then we said $\|\Pi x\|^2 = \sigma^T A_x^T A_x \sigma$, which was a quadratic form.

We shall write $\Pi_{ij} = \frac{\sigma_{ij} \delta_{ij}}{\sqrt{s}}$ where $\delta_{ij} \in \{0, 1\}$ is a random variable indicating whether the corresponding entry of Π was chosen to be non-zero. (So the δ_{ij} are not independent.) For every $r \in [m]$, define $x(r)$ by $(x(r))_i = \delta_{ri} x_i$. The claim is now that $\Pi x = A_x \sigma$ where A_x is an $m \times mn$ matrix whose ith row contains $x(r)^T/\sqrt{s}$ in the ith block of size n and zeros elsewhere. Using the inequality, we observe that $A_x^T A_x$ is a block-diagonal matrix as before. And since we're bounding the difference between $\sigma^T A_x^T A_x \sigma$ and its expectation, it is equivalent to bound $\sigma^T B_x \sigma$ where B_x is $A_x^T A_x$ with its diagonals zeroed out.

Now condition on B_x and recall that the inequality says that for all $p \geq 1$, $\|\sigma^T B_x \sigma\|_p \leq p\|B_x\| + \sqrt{p}\|B_x\|_F$. Then, taking p-norms with respect to the δ_{ij}

and using the triangle inequality, we obtain the bound

$$\|\sigma^T B_x \sigma\|_p \leq p\|\|B_x\|\|_p + \sqrt{p}\|\|B_x\|_F\|_p$$

If we can bound the right-hand-side, we will obtain required DJL result by application of Markov's inequality, since $\sigma^T B_x \sigma$ is positive. Therefore, it suffices to bound the p-norms with respect to the δ_{ij} of the operator and Frobenius norms of B_x.

Since B_x is block-diagonal and its ith block is $x(r)x(r)^T - \Lambda(r)$ where $\Lambda(r)$ is the diagonal of $x(r)x(r)^T$, we have $\|B_x\| = \frac{1}{s}\max_{1 \leq r \leq m}\|x(r)x(r)^T - \Lambda(r)\|$. But the operator norm of the difference of positive-definite matrices is at most the max of either operator norm. Since both matrices have operator norm at most 1, this concludes $\|B_x\| \leq 1/s$ always.

See that we defined $B_x = A_x^T A_x$ as the center of the product from before, but with the diagonals zeroed out. B_x is a block-diagonal matrix with m blocks $B_{x,1}, \ldots, B_{x,r}$ with

$$(B_{x,r})_{i,j} = \begin{cases} 0, & i = j \\ \delta_{r,i}\delta_{r,j}x_ix_j, & i \neq j. \end{cases}$$

We can state the Frobenius norm as

$$\|B_x\|_F^2 = \frac{1}{s^2}\sum_{r=1}^{m}\sum_{i \neq j}\delta_{r_i}\delta_{r_j}x_i^2x_j^2$$

$$= \frac{1}{s^2}\sum x_i^2x_j^2\left(\sum_{r=1}^{m}\delta_{r_i}\delta_{r_j}\right)$$

where we define the expression in the parentheses to be W_{ij}.

Claim

$$\|W_{ij}\|_p \lesssim p$$

Let us assume the claim and show that the Frobenius norm is correct.

$$\|\|B_x\|_F\|_p = (e[\|B_x\|_F]^p)^{1/p}$$

$$= (((e[\|B_x\|_F]^2)^{p/2})^{2/p})^{1/2}$$

$$= \|\|B_x\|_F^2\|_{p/2}^{1/2}$$

$$\leq \|\|B_x\|_F^2\|_p^{1/2}$$

$$= \left(\|\frac{1}{s^2}\sum_{i \neq j}x_i^2x_j^2W_{ij}\|_p\right)^{1/2}$$

$$\leq \frac{1}{s}\|\sum_{i \neq j}x_i^2x_j^2W_{ij}\|_p^{1/2}$$

$$\lesssim \frac{\sqrt{p}}{s} \left(\sum_{i \neq j} x_i^2 x_j^2 \right)^{1/2}$$

$$\leq \frac{\sqrt{p}}{s}$$

$$\simeq \frac{\varepsilon}{\sqrt{\ln 1/\delta}} \simeq \frac{1}{\sqrt{m}}$$

Now,

$$\| \|\Pi x\|_2^2 - 1\|_p = \|\sigma^T B_x \sigma\|_p \leq \sqrt{\frac{p}{m}} + \frac{p}{s}$$

$$(Markov) \implies \quad \mathbb{P}(| \|\Pi x\|^2 - 1| > \varepsilon) \leq \frac{\| \|\Pi x\|_2^2 - 1\|_p^p}{\varepsilon^p}$$

$$\leq 2^p \cdot \left(\frac{max(\sqrt{\frac{p}{m}}, \frac{p}{s})}{\varepsilon} \right)^p < \delta$$

Proof Let us just fix column i. It has s nonzero elements somewhere. There's another column j, and the question is how many of the nonzero locations of i match with nonzero elements of j. Let's have Y_t be an indicator random variable for column j having a nonzero element in the tth nonzero row of i. Then $W_{ij} = \sum_{t=1}^{s} Y_t$.

However, the moments are dominated by the independent case.

$$e\left[\sum_t Y_t \right]^p = \sum_{s=1}^{min(p,s)} \sum_{d_1,d_2,...,d_l \sum d_j = p} \sum_{i_1 < i_2 < \cdots < i_l} e\left[\prod_{q=1}^{s} Y_{i_q} \right]$$

The expected value of any Y_t is s/n. The product at the end is just $(s/n)^l$ in the independent case. Here, it is a conditional product,

$$e\left[\prod_{q=1}^{l} Y_l \right] = \mathbb{P}(Y_{i_1} = 1) \cdot \mathbb{P}(Y_{i_2} = 1 | Y_{i_1} = 1) \cdots$$

$$= \frac{s}{m} \cdot \frac{s-1}{m-1} \cdots \frac{s-l+1}{m-l+1}$$

$$\leq (s/m)^l$$

So the sum is dominated by the independent case, which can be handled via Bernstein's inequality.

The runtime to apply the sparse JL map is $O(s \times supp(x))$.

2.8 Fast Johnson–Lindenstrauss Transform

We take another approach that will give $O(n \lg n)$ time, which is better in cases where x is dense. This is based on Ailon and Chazelle (Ailon and Chazelle 2009) and is called as the Fast Johnson Lindenstrauss Transform (FJLT). The original construction in this area is due to Ailon and Chazelle (though fast Fourier ideas have obviously existed much longer), but there are many others. Theoretically, they are all to a first approximation the same; but practically, there can be a big difference between them.

Here is the definition of Π:

$$\Pi = \frac{1}{\sqrt{m}} \cdot PHD$$

where P is an $m \times n$ sampling matrix (very sparse matrix in expectation, only a fraction of the elements are nonzero). H is $\sqrt{n}$ times an orthogonal $n \times n$ matrix, i.e. $H^T H = n \cdot I$. Also $max|H_{ij}| = O(1)$, and computing Hx should be fast for any x. D is an $n \times n$ diagonal matrix with random signs $\alpha_1, \ldots, \alpha_n$ along the diagonal. So, HD is an orthogonal matrix, meaning in particular that the Euclidean norm of vectors to which it is applied does not change.

We will let $P = S_\eta$ be an $n \times n$ diagonal matrix where the ith diagonal entry η_i equals 1 with probability m/n and 0 otherwise, and the η_i are independent across i. See that an example of H could be the unnormalized discrete Fourier transform. Another possibility for H is the unnormalized Hadamard matrix where $H_{i,j} = (-1)^{\langle i,j \rangle}$. Here n is a power of 2 and we are interpreting i, j as elements of $\mathbb{F}_2^{\log_2 n}$. Both of these matrices allow Hx to be computed in time $O(n \log n)$. In general, $n \times n$ matrices F which are orthogonal with $max_{i,j} |F_{i,j}| = O(1/\sqrt{n})$ are called *bounded orthonormal systems*.

In (Ailon and Chazelle 2009) the following is shown. The key ingredient of their argument is the well-known Khintchine inequality from functional analysis.

Claim

$$\forall x, \|x\|_2 = 1, \mathbb{P}_\alpha \left(\|HDx\|_\infty > c \cdot \sqrt{\frac{\lg(n/\delta)}{n}} \right) < \delta/2$$

If we restrict α so that the above claim holds, then Bernstein implies that for

$$m \geq \frac{\log(1/\delta) \log(n/\delta)}{\varepsilon^2},$$

we will have $\|(1/\sqrt{m}) PHDx\|_2^2 = (1 \pm \varepsilon)\|x\|_2^2$ with probability $1 - \delta/2$. Thus by a union bound, the overall failure probability is δ.

Suppose we want to have $O(\varepsilon^{-2} \log(1/\delta))$ rows, we can do this by using the matrix $\Pi' \cdot (1/\sqrt{m}) PHD$, where Π' is for example a dense random sign matrix with $m' = O(\varepsilon^{-2} \log(1/\delta))$ rows.

The total time to apply $\Pi' \cdot \Pi$ is then $O(n \log n + m' \cdot m)$.

Rather different analysis can improve the $\log(n/\delta)$ dependence in m to be $\log(m/\delta)$ as follows.

Theorem 2.10 *Let $x \in \mathbb{R}^n$ be an arbitrary unit norm vector, and suppose $0 < \varepsilon, \delta < 1/2$. Also let $\Pi = S_n H D$ as described above with a number of rows equal to $m \gtrsim \varepsilon^{-2} \log(1/\delta) \log(1/(\varepsilon\delta))$. Then*

$$\mathbb{P}_{\Pi}(|\,\|\Pi x\|_2^2 - 1| > \varepsilon) < \delta.$$

Proof Let η' be an independent copy of η, and let $\sigma \in \{-1, 1\}^n$ be uniformly random. Write $z = HDx$ so that $\|\Pi x\|_2^2 = \sum_i \eta_i z_i^2$.

$$\|\frac{1}{m} \sum_{i=1}^n \eta_i z_i^2 - 1\|_p = \|\|\frac{1}{m} \sum_i \eta_i z_i^2 - 1\|_{L^p(\eta)}\|_{L^p(\alpha)} \tag{2.24}$$

$$= \|\|\frac{1}{m} \sum_i \eta_i z_i^2 - \frac{1}{m} \mathrm{e}_{\eta'} \sum_i \eta_i' z_i^2\|_{L^p(\eta)}\|_{L^p(\alpha)}$$

$$\leq \|\|\frac{1}{m} \sum_i z_i^2 (\eta_i - \eta_i')\|_{L^p(\eta,\eta')}\|_{L^p(\alpha)} \text{ (Jensen)}$$

$$= \|\|\frac{1}{m} \sum_i \sigma_i z_i^2 (\eta_i - \eta_i')\|_{L^p(\eta,\eta')}\|_{L^p(\alpha)} \text{ (equal in distribution)}$$

$$\leq \frac{2}{m} \cdot \|\|\sum_i \sigma_i \eta_i z_i^2\|_{L^p(\eta)}\|_{L^p(\alpha)} \text{ (triangle inequality)}$$

$$\leq \frac{2}{m} \cdot \|\sum_i \sigma_i \eta_i z_i^2\|_p$$

$$\lesssim \frac{\sqrt{p}}{m} \cdot \|\left(\sum_i \eta_i z_i^4\right)^{1/2}\|_p \text{ (Khintchine)}$$

$$\leq \frac{\sqrt{p}}{m} \cdot \|(\max_i \eta_i |z_i|) \cdot \left(\sum_i \eta_i z_i^2\right)^{1/2}\|_p$$

$$\leq \frac{\sqrt{p}}{m} \cdot \|\max_i \eta_i z_i^2\|_p^{1/2} \cdot \|\sum_i \eta_i z_i^2\|_p^{1/2} \text{ (Cauchy-Schwarz)}$$

$$\leq \sqrt{\frac{p}{m}} \cdot \|\max_i \eta_i z_i^2\|_p^{1/2} \cdot \left(\|\frac{1}{m} \sum_i \eta_i z_i^2 - 1\|_p^{1/2} + 1\right) \text{ (triangle inequality)}$$

$$\tag{2.25}$$

We will now bound $\|\max_i \eta_i z_i^2\|_p^{1/2}$. Define $q = \max\{p, \log m\}$ and see $\|\cdot\|_p \leq \|\cdot\|_q$. Then

$$\|\max_i \eta_i z_i^2\|_q = \left(\mathrm{e}_{\alpha,\eta} \max_i \eta_i z_i^{2q}\right)^{1/q}$$

$$\leq \left(e_{\alpha,\eta} \sum_i \eta_i z_i^{2q} \right)^{1/q}$$

$$= \left(\sum_i e_{\alpha,\eta} \eta_i z_i^{2q} \right)^{1/q}$$

$$\leq \left(n \cdot \max_i e_{\alpha,\eta} \eta_i z_i^{2q} \right)^{1/q}$$

$$= \left(n \cdot \max_i (e_\eta \eta_i) \cdot (e_\alpha z_i^{2q}) \right)^{1/q} \quad (\alpha, \eta \text{ independent})$$

$$= \left(m \cdot \max_i e_\alpha z_i^{2q} \right)^{1/q}$$

$$\leq 2 \cdot \max_i \|z_i^2\|_q \quad (m^{1/q} \leq 2 \text{ by choice of } q)$$

$$= 2 \cdot \max_i \|z_i\|_{2q}^2$$

$$\lesssim q \quad \text{(Khintchine)} \tag{2.26}$$

H is an unnormalized bounded orthonormal system.

Establishing $E = \|\frac{1}{m} \sum_i \eta_i z_i^2 - 1\|_p^{1/2}$ and integrating above equations, we find that for some constant $C > 0$

$$E^2 - C\sqrt{\frac{pq}{m}} E - C\sqrt{\frac{pq}{m}} \leq 0,$$

implying $E^2 \lesssim \max\{\sqrt{pq/m}, pq/m\}$. By the Markov inequality

$$\mathbb{P}(|\|\Pi x\|_2^2 - 1| > \varepsilon) \leq \varepsilon^{-p} \cdot E^{2p},$$

and thus to achieve the theorem statement it suffices to set $p = \log(1/\delta)$ then choose $m \gtrsim \varepsilon^{-2} \log(1/\delta) \log(m/\delta)$.

Remark 2.3 The Fast Johnson Lindenstrauss Transform gives suboptimal m. For necessary optimal m, one can use the embedding matrix $\Pi'\Pi$, where Π is the FJLT and Π' is, say, a dense matrix with Rademacher entries having the optimal $m' = O(\varepsilon^{-2} \log(1/\delta))$ rows. In (Ailon and Chazelle 2009), this term enhanced by replacing the matrix S with a random sparse matrix P.

Remark 2.4 The analysis for the FJLT, such as the approach in (Ailon and Chazelle 2009), would achieve a bound on m of $O(\varepsilon^{-2} \log(1/\delta) \log(n/\delta))$. Such analyses operate by, using the notation of the proof of Theorem 2.10, first conditioning on $\|z\|_\infty \lesssim \sqrt{\log(n/\delta)}$, then completing the proof using Bernstein's inequality.

2.9   Sublinear-Time Algorithms: An Example

In this example, we discuss a type of approximation that makes sense for outputs of decision problems.

Example 2.1   sequence monotonicity, version 1 Given an ordered list $X_1, \ldots, X_n$ of elements (with partial order '$\leq$' on them), the list is $X_1 \leq \cdots \leq X_n$.

Instead of looking at each single sequence element, we consider the following version,

Example 2.2   sequence monotonicity, version 2.

Given an ordered list $X_1, \ldots, X_n$ of elements (with partial order '$\leq$' on them) and a real fraction $\varepsilon \in [0, 1]$, the list is *close to monotone*. That means, a list is ε-*close to monotone* if it has a monotone subsequence of length $(1 - \varepsilon)n$.

If the list is monotone, the test should pass with probability $3/4$. If the list is ε-far from monotone, the test should fail with probability $3/4$.

Remark 2.5   The choice of $3/4$ is arbitrary; any constant bounded away from $1/2$ works equally well. We can expand the definition from our constant to a different constant $1 - \beta$ by repeating the algorithm $O(\log \frac{1}{\beta})$ times and taking the majority answer.

Remark 2.6   The behaviour of the test on inputs that are very close to monotone, but are not monotone, is undefined. (Those inputs are ε'-close with $0 \leq \varepsilon' \leq \varepsilon$.)

We present some cases below:
Select $i < j$ randomly and test $x_i < x_j$.
We will show that complexity of such case is $\Omega(\sqrt{n})$.
For constant c:

$$\underbrace{c, c - 1, \ldots, 1}, \underbrace{2c, 2c - 1, \ldots, c + 1}, \ldots, \ldots, \ldots, \underbrace{n, n - 1, \ldots, n - c + 1}.$$

The longest monotone subsequence has length n/c rather small, hence we expect this sequence to fail the test. Interestingly the test passes when it selects i, j from different groups.

If the test is repeated by repeatedly selecting new pairs i, j, each time discarding the old pair, and checking each such pair independently of the others, then $\Omega(n)$ pairs are needed. However, if the test is repeated by selecting k indices and testing whether the subsequence induced by them is monotone, then $\Theta(\sqrt{n/c})$ samples are required using the Birthday Paradox. The Birthday Paradox states that in a random group of people, there is about a 50 % chance that two people have the same birthday. There are many reasons why this seems like a paradox.

In other way, select i randomly and test $x_i \leq x_{i+1}$. For some constant c, and consider the following sequence (of n elements):

$$\underbrace{1, 2, \ldots, n/c}, \underbrace{1, 2, \ldots, n/c}, \ldots, \underbrace{1, 2, \ldots, n/c}.$$

Now, the longest monotone subsequence has length $c + \frac{n}{c} - 1$ — rather small, therefore we expect this sequence to fail the test. However, the test passes unless the i it selects is a *border point*, which happens with probability c/n.

Therefore we expect to have a linear number of samples before detecting an input that should be rejected.

This would check that the sequence is locally monotone, and also monotone at large distances, but would not verify that it is monotone in middle-range gaps. And counter-examples can be found. However, there exists a correct, $O(\log n)$-samples algorithms that works by testing pairs at different distances $1, 2, 4, 8, \ldots, 2^k, \ldots, n/2$.

Lemma 2.9 X_i *are pairwise distinct.*

Proof Replace each X_i by the tuple (X_i, i) and use dictionary order to compare '(X_i, i)' to '(X_j, j)', compare the first coordinate and use the second coordinate to break ties.

Remark 2.7 This move does not demand the sublinearity of the algorithm because it does not require any pre-processing; the transformation can be done on the fly as each element is examined and compared.

'$[n]$' indicates the set $\{1, 2, \ldots, n\}$ of positive integers.
'$\in_R$' indicates assignment of a random member of the set on its right hand side (RHS) to the variable on its left hand side (LHS). If the distribution is not given, it is the uniform distribution.

For example, '$x \in_R [3]$' assigns to x one of the three smallest positive integers, chosen uniformly.

The procedure is:

- Repeat $O(1/\varepsilon)$ times:

  - Select $i \in_R [n]$
  - Query (obtain) the value X_i
  - Do binary search for X_i
  - If either
       An inconsistency was found during the binary search;

       X_i was not found;
    then return **fail**.
  - Return **pass**.

During the binary search, we maintain an interval of allowed values for the next value we query. The interval begins as $[-\infty, +\infty]$. The upper and lower bounds are updated whenever we take a step to the left or to the right, respectively. Whenever we query a value we state that it is in the interval and raise an inconsistency if it is not.

This algorithm's time complexity is $O(\frac{1}{\varepsilon} \log n)$, since the augmented binary search and the choosing of a random index cost $O(\log n)$ steps each; and those are repeated $O(1/\varepsilon)$ times. We will now show that the algorithm satisfies the required behavior. We will define which indices are 'good' and relate the number of bad indices to the length of a monotone sequence of elements at good indices.

Definition 2.4 An index i is *good* if augmented binary search for i is accomplished.

Remark 2.8 If $\geq \varepsilon n$ indices are bad, then Prob[pass] $< 1/4$.

Proof Let c be the constant under the '$O(1/\varepsilon)$ repetitions' clause. Then

$$\text{Prob[pass]} \leq (1 - \varepsilon)^{c/\varepsilon} < \frac{1}{4}, \tag{2.27}$$

where the last inequality follows by setting c to a large enough constant value.

Theorem 2.11 *The algorithm has 2-sided error less than one quarter. It accepts good inputs with probability 1 and rejects bad inputs with probability at least 3/4.*

Proof When the list is monotone, it passes with trust since the binary search works and the X_i are considered distinct. It needs to show that "far from monotone" lists are rejected with high likelihood.

Suppose that an input passes with probability $> 1/4$, we shall prove that it is ε-close.

Let $X_1, \ldots, X_n$ be received with probability $> 1/4$. By Eq. (2.27), the number of bad indices is $< \varepsilon n$. Hence $\geq (1 - \varepsilon)n$ indices are good.

Claim *Suppose we delete every element at bad indices, the remaining sequence is monotone.*

Proof Let $i < j$ be two indices. Consider the paths in the binary-search tree from the root to i and to j. These two paths have longest prefix common. Then it is enough to prove that $x_i \leq z \leq x_j$.

When the path to x_i is a prefix of the path to x_j, then $x_i = z$. Alternatively x_i is a descendant of a z's left or right child. Since i is good, then x_i must be a descendant of z's left child; for the same reason x_i must be smaller than z. Thus, $x_i \leq z$ always. By symmetry, $z \leq x_j$. Hence $x_i \leq x_j$.

Hence proved.

2.10 Minimum Spanning Tree

Let us consider a connected undirected graph $G = (V, E)$ where the degree of each vertex is at most d. In addition, each edge (i, j) has an integer weight $w_{ij} \in [w] \cup$

{∞}. The graph is given in an adjacency list format, and edges of weight ∞ do not appear in it. The aim is to find the weight of a minimum spanning tree (MST) of G. Specifically, if we let $w(T) = \sum_{(ij)\in T} w_{ij}$ for $T \subseteq E$, then our objective is to find

$$M = \min_{T \text{ spans } G} w(T) .$$

Our objective is to select a subset of the edges of minimum total length such that all the vertices are connected. It is immediate that the resulting set of edges forms a spanning tree every vertex must be included; Cycles do not improve connectivity and only increase the total length. Therefore, the problem is to find a spanning tree of minimum total length.

There are many greedy procedures work for this problem. One can either start with the empty graph and consecutively add edges while avoiding forming cycles, or start with the complete graph and consecutively remove edges while maintaining connectivity. The crucial aspect is the order in which edges are considered for addition or deletion. We present three basic greedy procedures in the following lines, all of which lead to optimal tree constructions:

- *Kruskals algorithm*: Consider edges in increasing order of length, and pick each edge that does not form a cycle with previously included edges.
- *Prims algorithm*: Start with an arbitrary node and call it the root component; at every step, grow the root component by adding to it the shortest edge that has exactly one end-point in the component.
- *Reverse delete*: Start with the entire graph, and consider edges for deletion in order of decreasing lengths. Remove an edge as long as the deletion does not disconnect the graph.

We now consider a fundamental algorithm for finding the MST, which proceeds in $O(\log n)$ phases. In each iteration, the minimum weight edge on each vertex is added and the resulting connected components are collapsed to form new vertices. This algorithm can be implemented in the dynamic stream setting in $O(\log^2 n)$ passes by imitating each iteration in $O(\log n)$ passes of the dynamic graph stream. In the first pass, we ρ_0-sample an incident edge on each vertex without considering the weights. Suppose we sample an edge with weight w_v on vertex v. In the next pass, we repeat ρ_0 sample incident edges but we ignore all edges of weight at least w_v on vertex v when we create the sketch. Repeating this process $O(\log n)$ assures that we succeed in finding the minimum weight edge incident on each vertex. Hence the algorithm takes $O(\log^2 n)$ passes as claimed.

Since we are interested in sub-linear time algorithms for this problem, and therefore, cannot hope to find M, we focus on finding an ε-multiplicative estimate of M, that is, a weight $\hat{M}$ which satisfies

$$(1 - \varepsilon)M \le \hat{M} \le (1 + \varepsilon)M .$$

We see that $n - 1 \leq M \leq w \cdot (n - 1)$, where $n = |V|$. This follows since G is connected, and thus, any spanning tree of it consists of $n - 1$ edges, and by the premise on the input weights.

In what follows, we relate the weight of a MST of G to the number of connected components in certain subgraphs of G. We begin by introducing the following notation for a graph G:

Let $G^{(i)} = (V, E^{(i)})$ be the subgraph of G that consists of the edges having a weight of at most i.

Let $C^{(i)}$ be the number of connected components in $G^{(i)}$.

Let us consider two simple cases. The first case is when $w = 1$, namely, all the edges of G have a weight of 1. In this case, it is clear that the weight of a MST is $n - 1$. Now, let us consider the case that $w = 2$, and let us focus on $G^{(1)}$. Clearly, one has to use $C^{(1)} - 1$ edges (of weight 2) to connect the connected components in $G^{(1)}$. This implies that the weight of a MST in this case is

$$2 \cdot (C^{(1)} - 1) + 1 \cdot (n - 1 - (C^{(1)} - 1)) = n - 2 + C^{(1)}.$$

We extend and formalize the intuition presented above. Specifically, we characterize the weight of a MST of G using the $C^{(i)}$'s, for any integer w.

Claim $M = n - w + \sum_{i=1}^{w-1} C^{(i)}$.

Proof Let α_i be the number of edges of weight i in any MST of G. Obviously, it is well-known that all minimum spanning trees of G have the same number of edges of weight i, and hence, the α_i's are well defined. It is easy to validate that the number of edges having weight greater than ℓ is equal to the number of connected components in $G^{(\ell)}$ minus 1. That is, $\sum_{i=\ell+1}^{w} \alpha_i = C^{(\ell)} - 1$, where $C^{(0)}$ is set to be n. Therefore

$$
\begin{aligned}
M &= \sum_{i=1}^{w} i \cdot \alpha_i \\
&= \sum_{i=1}^{w} \alpha_i + \sum_{i=2}^{w} \alpha_i + \sum_{i=3}^{w} \alpha_i + \cdots + \alpha_w \\
&= (n - 1) + (C^{(1)} - 1) + (C^{(2)} - 1) + \cdots + (C^{(w-1)} - 1) \\
&= n - w + \sum_{i=1}^{w-1} C^{(i)}
\end{aligned}
$$

2.10.1 Approximation Algorithm

Algorithm MST_approx, formally defined below, estimates the weight of the MST.

$MST\_approx(G, \varepsilon, w)$
    for $i = 1$ to $w - 1$

$$\hat{C}^{(i)} = approx\_num\_cc(G^{(i)}, \varepsilon/(2w))$$
output $\hat{M} = n - w + \sum_{i=1}^{w-1} C^{(i)}$

See that there are w calls to $approx\_num\_cc$. Recall that the running time of this procedure is $O(d/(\varepsilon/(2w))^4) = O(dw^4/\varepsilon^4)$, and hence, the running time of MST_approx is $O(dw^5/\varepsilon^4)$. It is worth noting that rather than extracting $G^{(i)}$ from G for each call of $approx\_num\_cc$, that makes the algorithm non-sublinear time, we simply modify $approx\_num\_cc$ so it ignores edges with weight greater than i.

We establish that $(1 - \varepsilon)M \leq \hat{M} \leq (1 + \varepsilon)M$ with high probability. For this purpose, recall that $approx\_num\_cc$ outputs an estimation $\hat{C}^{(i)}$ of the number of connected components which satisfies $|\hat{C}^{(i)} - C^{(i)}| \leq n\varepsilon/(2w)$ whp. Consequently, we get that $|M - \hat{M}| \leq n\varepsilon/2$ whp. Notice that $M \geq n - 1 \geq n/2$, where the last inequality is valid for any n, i.e., $n \geq 2$. Therefore, $|M - \hat{M}| \leq M\varepsilon$, which completes the proof.

The modern procedure for finding an ε-multiplicative estimate of M has a running time of $O(dw/\varepsilon^2 \cdot \log dw/\varepsilon)$. On the lower bound side, it is known that the running time of any algorithm must be $\Omega(dw/\varepsilon^2)$.

Chapter 3
Linear Algebraic Models

3.1 Introduction

This chapter presents some of the fundamental linear algebraic tools for large scale
data analysis and machine learning. Specifically, the focus will fall on large scale
linear algebra, including iterative, approximate and randomized algorithms for basic
linear algebra computations and matrix functions. In the last decade, several algorithms for numerical linear algebra have been proposed, with substantial gains in
performance over older algorithms (Nelson et al. 2014; Nelson 2015). Algorithms
of such type are mostly pass-efficient, requiring only a constant number of passes
over the matrix data for creating samples or sketches, and other work. Most these
algorithms require at least two passes for their efficient performance guarantees,
with respect to error or failure probability. Such a one-pass algorithm is close to the
streaming model of computation, where there is one pass over the data, and resource
bounds are sublinear in the data size.

Definition 3.1 $\Pi \in \mathbb{R}^{m \times n}$ and D is a distribution over Π satisfies the (ε, δ, p)−JL
moment property if for any $x \in S^{n-1}$ we have $\mathbb{E}_{\Pi \sim D}|\|\Pi x\|_2^2 - 1|^p < \varepsilon^p \delta$

Example 3.1 1. $\Pi_{ij} = \pm 1/\sqrt{m}$. This induces $(\varepsilon, \delta, 2)-$ JL moment property with
  $m \geq 1/\varepsilon^2 \delta$ and $(\varepsilon, \delta, \log(1/\delta))-$ JL moment property with $m \geq \log(1/\delta)/\varepsilon^2$
2. We have $(\varepsilon, \delta, 2)-$ JL moment property with $m \geq 1/\varepsilon^2 \delta$

Claim *Suppose Π comes from $(\varepsilon, \delta, p)-$ JL moment property for some $p \geq 2$. Then
for any A, B with n rows, we have*

$$\mathbb{P}_{\Pi \sim D}(\|A^T B - (\Pi A)^T (\Pi B)\|_F > \varepsilon \|A\|_F \|B\|_F) < \delta \qquad (3.1)$$

Proof The proof is left as an exercise for the reader.

Definition 3.2 Given $E \subset \mathbb{R}^n$ a linear subspace, $\Pi \in \mathbb{R}^{m \times n}$ is an ε-**subspace
embedding** for E if

© The Author(s), under exclusive license to Springer Nature Switzerland AG 2018   65
R. Akerkar, *Models of Computation for Big Data*, SpringerBriefs in Advanced
Information and Knowledge Processing, https://doi.org/10.1007/978-3-319-91851-8_3

$$\forall z \in E : (1 - \varepsilon)\|z\|_2^2 \leq \|\Pi z\|_2^2 \leq (1 + \varepsilon)\|z\|_2^2$$

We can frame these subspace embeddings to the approximate matrix multiplication methods:

- $E = \text{colspace}(A) \implies \forall x \in R^d : (1 - \varepsilon)\|Ax\|_2^2 \leq \|\Pi Ax\|_2^2 \leq (1 + \varepsilon)\|Ax\|_2^2$
- $C = D^T D = (\Pi A)^T (\Pi A)$.
  The above statement $\implies \forall x \in R^d : |x^T[A^T A - (\Pi A)^T (\Pi A)]x| \leq \varepsilon\|Ax\|_2^2$.
  This is a better bound, which preserves x.

As we know, any linear subspace is the column space of some matrix. Thus, we will represent them as matrices.

Claim *For any A of rank d, there exists a 0-subspace embedding* $\Pi \in \mathbb{R}^{d \times n}$ *with* $m = d$, *but no ε-subspace embedding* $\Pi \in \mathbb{R}^{m \times n}$ *with $\varepsilon < 1$ if $m < d$.*

Proof Now, let us imagine that there is an ε-subspace embedding $\Pi \in \mathbb{R}^{m \times n}$ for $m < d$. Then, the map $\Pi : E \to \mathbb{R}^m$ has a non-trivial kernel. Actually, there is some $x \neq 0$ such that $\Pi x = 0$. However, $\|\Pi x\|_2 \geq (1 - \varepsilon)\|x\|_2 > 0$ is a contradiction. For the $m \leq d$ case, begin by rotating the subspace E to become $\text{span}(e_1, \ldots, e_d)$ via multiplication by an orthogonal matrix, and then project to the first d coordinates.

Theorem 3.1 (Singular value decomposition) *Every $A \in \mathbb{R}^{n \times d}$ of rank r has a singular value decomposition (SVD)*

$$A = U \Sigma V^T$$

where $U \in \mathbb{R}^{n \times r}$ has orthonormal columns, $r = \text{rank}(A)$, $\Sigma \in \mathbb{R}^{r \times r}$ is diagonal with strictly positive entries σ_i on the diagonal, and $V \in \mathbb{R}^{d \times r}$ has orthonormal columns so $U^T U = I$ and $V^T V = I$

When we have $U \Sigma V^T$, we can set $\Pi = U^T$. There are procedures to compute U, Σ, V^T in $O(nd^2)$ time:

Theorem 3.2 (Demmel et al. 2007) *We can approximate SVD well in time $\tilde{O}(nd^{\omega-1})$ where ω is the constant in the exponent of the complexity of matrix multiplication. Here the tilde hides logarithmic factors in n.*

Definition 3.3 Suppose we are given $A \in \mathbb{R}^{n \times d}$, $b \in \mathbb{R}^n$ where $n \gg d$. We want to solve $Ax = b$; however, since the system is over-constrained, an exact solution does not exist in general. In the *least squares regression* (LSR) problem, we instead want to solve the equation in a specific approximate sense: we want to compute

$$x^{LS} = \text{argmin}_{x \in \mathbb{R}^d} \|Ax - b\|_2^2$$

The choice of the function to be optimized is not arbitrary. For example, assume that we have some system, and one of its parameters is a linear function of d other parameters. Actually, we experimentally observe a linear function plus some random error. Under certain premises, errors have mean 0, same variance, and are independent, then least squares regression is provably the best estimator out of a certain class of estimators.

Now consider that $\{Ax : x \in \mathbb{R}^d\}$ is the column span of A. Part of b lives in this column space, and part of it lives orthogonally. Then, the x^{LS} we need is the projection of b on that column span. Let $A = U\Sigma V^T$ be the SVD of A. Then the projection of b satisfies

$$\text{Proj}_{\text{Col}(A)} b = UU^T b$$

hence we can set $x^{LS} = V\Sigma^{-1}U^T b = (A^T A)^{-1}A^T b$. Then we have $Ax^{LS} = U\Sigma V^T V\Sigma^{-1}U^T b = UU^T b$. Thus, we can solve LSR in $O(nd^2)$ time.

Claim *If* $||Dx||^2 = (1+\varepsilon)||A'x||^2$ *for all* x $A' = [A|b]$ *then if* $\tilde{x}^{LS} = argmin_{x'=[x|-1]}||Dx||_2^2$, *then:*

$$||Ax - b||_2^2 \leq \frac{1+\varepsilon}{1-\epsilon}||Ax^{LS} - b||_2^2$$

We are going to replace A with ΠA. Then we just need the SVD of ΠA, which only takes us $O(md^2)$ time. If m is like d then this is faster. However, we still need to find Π and apply it.

Claim *If* Π *is* ε-s.e. *for* $span(cols(A, b))$ *then if* $\tilde{x}^{LS} = argmin_{|\Pi Ax - \Pi b|_2^2}$, *then:*

$$||A\tilde{x}^{LS} - b||_2^2 \leq \frac{1+\varepsilon}{1-\epsilon}||Ax^{LS} - b||_2^2$$

Proof $||\Pi A\tilde{x}^{LS} - \Pi b||_2^2 \leq ||\Pi A\tilde{x}^{LS} - \Pi b||_2^2 \leq (1+\varepsilon)||Ax^{LS} - b_2^2||$. Similarly for the left side of the inequality.

The total time to find $\tilde{x}^{LS}$ includes the time to find Π, the time to compute ΠA, Πb, and $O(md^2)$ (the time to find the SVD for ΠA).

3.2 Sampling and Subspace Embeddings

As with approximate matrix multiplication, there are two possible methods we will examine: sampling, and a Johnson-Lindenstrauss (JL) method.

Let $\Pi \in \mathbb{R}^{n \times n}$ be a diagonal matrix with diagonal elements η_i. η_i is 1. If we sample the ith row i of A (which can be written as a_i^T), 0 otherwise. $e[\eta_i] = p_i$.

$$A^T A = \sum_{k=1}^{n} a_k a_k^T$$

$$(\Pi A)^T (\Pi A) = \sum_{k=1}^{n} \frac{\eta_k}{p_k} a_k a_k^T$$

$$e(\Pi A)^T (\Pi A) = \sum_{k=1}^{n} \frac{e[\eta_k]}{p_k} a_k a_k^T = A^T A$$

If we used the sampling approach for approximate matrix multiplication, we select proportional to the l_2 norm for each row and decide what p_k should be. The number of rows is non-deterministic.

Now let us try by intuitive way for the probabilities p_k:

If we do not want any p_k's to be 0 - then we just miss a row. Define $R_i = \sup_{x \in R^d} \frac{|a_i^T x|^2}{||Ax||_2^2}$. If we don't set $p_k \geq R_k$, it doesn't make sense. Look at the event that we did sample row i. Then

$$(\Pi A)^T (\Pi A) = \frac{1}{p_i} a_i a_i^T + \sum_{k \neq i}^{n} \frac{\eta_k}{p_k} a_k a_k^T$$

Pick x which achieves the sup in the definition of R_i. Then $x^T (\Pi A)^T (\Pi A) x = \frac{|a_i^T x|^2}{p_i} + \sum (\text{non-negative terms}) \geq \frac{|a_i^T x|^2}{p_i} = ||Ax||_2^2 \frac{R_i}{p_i}$ If $p_i < R_i/2$, the previous expression evaluates to $2||Ax||_2^2$, thus, we are guaranteed to mess up because there is some x which makes our error too big. Therefore, we need some $p_i > R_i/2$.

Definition 3.4 Given A, the ith *leverage score* l_i is $l_i = a_i^T (A^T A)^{-1} a_i$, if A has column rank.

Claim $l_i = R_i$

Proof See that both R_i and l_i are basis independent i.e. if M is square/invertible, then:

$l_i(A) = l_i(AM)$ and $R_i(A) = R_i(AM)$
$R_i(AM) = \sup_x \frac{|a_i^T M x|^2}{||AMx||_2^2} = \sup_y \frac{|a_i yt|}{||Ay||_2^2} = R_i(A)$

Choose M s.t. $\tilde{A} = AM$ has orthonormal columns: $M = V \Sigma^{-1}$. Then, wlog $A = \tilde{A} = AM$ and:

$R_i = \sup_x \frac{|a_i^T x|^2}{||x||_2^2}$

Which x achieves the sup in R_i?
The vector $||a_i||$ itself. Thus $R_i = ||a_i||_2^2$
$l_i = a_i^T (A^T A)^{-1} a_i = ||a_i||_2^2$.

Theorem 3.3 (Drineas et al. 2006) *If we choose* $p_i \geq min\{1, \frac{lg\,(d/\delta)||u_i||_2^2}{\varepsilon^2}\}$, *then* $\mathbb{P}(\Pi$ *is not* $\varepsilon - s.e.$ *for* $A) < \delta$.

See $||u_i||_2^2 = ||UU^Te_i||_2^2 = ||\text{Proj}_A e_i||_2^2 \leq ||e_i||_2^2 = 1$. So, none of the leverage scores can be bigger than 1, and they sum up to d. The minimum with 1 is needed to the multiplicative factor times the legepave score. We can analyse this using non-commutative khintchine.

Let us consider the analysis by non-commutative khintchine.

Definition 3.5 The *Schatten-p norm* of A for $1 \leq p \leq \infty$ is $||A||_{S_p} = l_p$-norm of singular values of A.

If A has rank $\leq d$, see that $||A||_{S_p} = \Theta(||A||) = ||A||_{S_\infty}$ for $p \geq \lg d$ (by Holder's inequality).

Theorem 3.4 (Lust-Piquard and Pisier 1991)

$$e\left(||\sum_i \sigma_i A_i||_{S_p}^p\right)^{1/p} \leq \sqrt{p} \max\left\{||\sum_i A_i^T A_i||_{S_{p/2}}^{1/2}, ||\sum_i A_i A_i^T||_{S_{p/2}}^{1/2}||\right\}$$

The total samples required is $\frac{d \lg (d/\delta)}{\varepsilon^2}$.

$||\Pi Ax||_2^2 = ||Ax||_2^2(1 \pm \varepsilon)$ for all x is the same as $||\Pi U\Sigma V^T y||_2^2(1 \pm \varepsilon)$ for all y. Call $\Sigma V^T y = x$. Thus we want $\forall x, ||\Pi U||_2^2 = (1 \pm \varepsilon)||Ux||_2^2 = (1 \pm \varepsilon)||x||_2^2$.

Therefore, want $\sup_{||x||_2=1} x^T[(\Pi U)^T(\Pi U) - I]x < \varepsilon$, i.e. $||(\Pi U)^T(\Pi U) - I|| < \varepsilon$:

The columns of U form an orthonormal basis for E.

We want $\forall x \in E, ||\Pi x||_2^2 = (1 + \varepsilon)||x||_2^2$ i.e. $\sup_{x \in E \cap S^{n-1}} |||\Pi x||_2^2 - 1| < \varepsilon$. From Gordon's theorem: If $\Pi_{i,j} = \pm 1/\sqrt{m}$ then suffices to have $m > g^2(T) + 1/\varepsilon^2$.

$$g(E \cap S^{n-1}) = e_{g \in \mathbb{R}^n} \sup_{||x||_2=1} \langle g, Ux \rangle$$

$$= e_{g \in \mathbb{R}^n} \sup_{||x|_2=1} \langle U^T g, x \rangle$$

$$= e_{g' \in \mathbb{R}^d} \sup_{||x|_2=1} \langle g', x \rangle$$

$$= ||g'||_2$$

$$\leq e(||g'||_2^2)^{1/2} = \sum_i (e(g_i')^2)^{1/2} = \sqrt{d}$$

Thus, if we take a random Gaussian matrix, by Gordon's theorem, it will preserve the subspace as long as it has at least $\frac{d}{\varepsilon^2}$ rows.

We want our Π to have few rows. We should be able to find Π immediately. Multiplication with A should be fast. The problem is ΠA takes time $O(mnd)$ using for loops, which takes time $> nd^2$. Hence, we want to use "fast Π".

Definition 3.6 An (ε, δ, d) *oblivious subspace embedding* is a distribution D over $\mathbb{R}^{m \times n}$ s.t. $\forall U \in \mathbb{R}^{n \times d}, U^T U = I$: $\mathbb{P}_{\Pi \sim D}(||(\Pi U)^T(\Pi U) - I|| > \varepsilon) < \delta$.

This distribution doesn't depend on A or U. The Gaussian matrix provides an oblivious subspace embedding, however, Sarlós approach with a fast JL matrix solves too.

For any d-dimensional subspace $E \in R^n$ there exists a set $T \subset E \cap S^{n-1}$ of size $O(1)^d$ such that if Π preserves every $x \in T$ up to $1 + O(\varepsilon)$ then Π preserves all of E up to $1 + \varepsilon$.

So what does this mean, if we have distributional JL than that automatically implies we have an oblivious subspace embedding. We would set the failure probability in JL to be $\frac{1}{O(1)^d}$ which by union bound gives us a failure probability of OSE of δ.

3.3 Non-commutative Khintchine Inequality

Non-commutative Khintchine inequality plays a vital role in the recent developments in non-commutative Functional Analysis, and in particular in Operator Space Theory. For Noncommutative Khintchine let $\|M\|_p = (e\|M\|_{S_p}^p)^{\frac{1}{p}}$ with $\sigma_1, \cdots, \sigma_n$ are $\{1, -1\}$ independent Bernoulli. Than

$$\left\| \sum_i \sigma_i A_i \right\|_p \le \sqrt{p} \max \left\{ \left\| \left(\sum_i A_i A_i^T \right)^{\frac{1}{2}} \right\|_p, \left\| \left(\sum A_i^T A_i \right)^{\frac{1}{2}} \right\|_p \right\}$$

To take the square root of a matrix just produce the singular value decomposition $U \Sigma V^T$ and take the square root of each of the singular values.

We take

$$P(\|(\Pi U)^T (\Pi U) - I\| > \varepsilon) < \delta.$$

We know the given expression is

$$P(\|(\Pi U)^T (\Pi U) - I\| > \varepsilon) < \frac{1}{\varepsilon^p} E \|(\Pi U)^T (\Pi U) - I\|^p \le \frac{C^p}{\varepsilon^p} E \|(\Pi U)^T (\Pi U) - I\|_{S_p}^p$$

We wish to bound $\|(\Pi U)^T (\Pi U) - I\|_p$ and we know

$$(\Pi U)^T (\Pi U) = \sum_i z_i z_i^T$$

where z_i is the i'th row of ΠU. This all implies

$$\|(\Pi U)^T (\Pi U) - I\|_p = \| \sum_i z_i z_i^T - E \sum y_i y_i^T \|_p$$

where $y_i \sim z_i$. Now we do the usual trick with proving Bernstein. By convexity we interchange the expectation with the norm and obtain

$$\le \| \sum_i (z_i z_i^T - y_i y_i^T) \|_p$$

which is just the usual symmetrization trick assuming row of Π are independent. Then we simplify

$$\leq 2\Big\|\sum_i \sigma_i z_i z_i^T\Big\|_{L^p(\sigma,z)} \leq \sqrt{p}\Big\|\Big(\sum_i \|z_i\|_2^2 z_i z_i^T\Big)^{\frac{1}{2}}\Big\|_p$$

The following was observed by Cohen, noncommutative khintchine can be applied to sparse JL

$$m \geq \frac{d\, polylog(\frac{1}{\delta})}{\varepsilon^2}, s \geq \frac{polylog(\frac{d}{\delta})}{\varepsilon^2}$$

but Cohen is able to obtain $m \geq \frac{d\log(\frac{d}{\delta})}{\varepsilon^2}, s \geq \frac{\log(\frac{d}{\delta})}{\varepsilon}$ for s containing dependent entries as opposed to independent entries. There is a conjecture that the multiplies in $d\log(\frac{d}{\delta})$ is basically an addition and has been useful in compressed sensing.

3.4 Iterative Algorithms

Some forms of randomization have been used for several years in linear algebra. For example, the starting vectors in Lanczos algorithms are always random. A few years ago, new uses of randomization have proposed such as random mixing and random sampling, which can be combined to form random projections. These ideas have been explored theoretically and have found use in some specialized applications (e.g., data mining), but they have had little influence on mainstream numerical linear algebra. Tygert and Rokhlin (Rokhlin and Tygert 2008) and Avron et al. (Avron et al. 2010) shaded light on using gradient descent.

Definition 3.7 For a matrix A, the *condition number* of A is the ratio of its largest and smallest singular values.

Let Π be a $1/4$ subspace embedding for the column span of A. Then let $\Pi A = U\Sigma V^T$ (SVD of ΠA). Let $R = V\Sigma^{-1}$. Then by orthonormality of U

$$\forall x : \|x\| = \|\Pi A R x\| = (1 \pm 1/4)\|A R x\|$$

which means $AR = \tilde{A}$ has a good condition number. Then algorithm is the following

1. Pick $x^{(0)}$ such that

$$\|\tilde{A}x^{(0)} - b\| \leq 1.1\|\tilde{A}x^* - b\|$$

By using reduction to subspace embeddings with ε being constant.
2. Iteratively let $x^{(i+1)} = x^{(i)} + \tilde{A}^T(b - \tilde{A}x^{(i)})$ until some $x^{(n)}$ is obtained.

We will discuss an analysis using (Clarkson and Woodruff 2013). Observe that

$$\tilde{A}(x^{(i+1)} - x^*) = \tilde{A}(x^{(i)} + \tilde{A}^T(b - \tilde{A}x^{(i)}) - x^*) = (\tilde{A} - \tilde{A}\tilde{A}^T\tilde{A})(x^{(i)} - x^*),$$

where the last equality follows by expanding the RHS. Obviously, all terms disappear except for $\tilde{A}\tilde{A}^T b$ versus $\tilde{A}\tilde{A}^T \tilde{A}x^*$, which are equal since x^* is the optimal vector. So, x^* is the projection of b onto the column span of $\tilde{A}$.

Now let $AR = U'\Sigma'V'^T$ in SVD, then

$$
\begin{aligned}
\|\tilde{A}(x^{(i+1)-x^*})\| &= \|(\tilde{A} - \tilde{A}\tilde{A}^T\tilde{A})(x^{(i)} - x^*)\| \\
&= \|U'(\Sigma' - \Sigma'^3)V'^T(x^{(i)} - x^*)\| \\
&= \|(I - \Sigma'^2)U'\Sigma'V'^T(x^{(i)} - x^*)\| \\
&\leq \|I - \Sigma'^2\| \cdot \|U'\Sigma'V'^T(x^{(i)} - x^*)\| \\
&= \|I - \Sigma'^2\| \cdot \|\tilde{A}(x^{(i)} - x^*)\| \\
&\leq \frac{1}{2} \cdot \|\tilde{A}(x^{(i)} - x^*)\|
\end{aligned}
$$

$\tilde{A}$ has a good condition number, thus $O(\log 1/\varepsilon)$ iterations suffice to bring down the error to ε. Further, in every iteration, we have to multiply by AR; multiplying by A can be done in time proportional to the number of nonzero entries of A, $\|A\|_0$, and multiplication by R in time proportional to d^2. Ultimately, the pertinent term in the time complexity is $\|A\|_0 \log(1/\varepsilon)$, in addition the time to find the SVD.

3.5 Sarlós Method

This method is proposed by Sarlós (Sarlós 2006), where he asked what space and time lower bounds can be proven for any pass-efficient approximate matrix product, regression, or SVD algorithm. Key applications of low-rank matrix approximation by SVD include recommender systems, information retrieval via Latent Semantic Indexing, Kleinberg's HITS algorithm for web search, clustering, and learning mixtures of distributions.

Let

$$
\begin{aligned}
x^* &= \text{argmin}\|Ax - b\| \\
\tilde{x}^* &= \text{argmin}\|\Pi Ax - \Pi b\|. \\
A &= U\Sigma V^T \text{ in SVD} \\
Ax^* &= U\alpha \text{ for } \alpha \in \mathbb{R}^d \\
Ax^* - b &= -w \\
A\tilde{x}^* - Ax^* &= U\beta
\end{aligned}
$$

Then, $OPT = \|w\| = \|Ax^* - b\|$. We have

$$\|A\tilde{x}^* - b\|^2 = \|A\tilde{x}^* - Ax^* + Ax^* - b\|^2$$
$$= \|A\tilde{x}^* - Ax^*\|^2 + \|Ax^* - b\|^2 \text{ (they are orthogonal)}$$
$$= \|A\tilde{x}^* - Ax^*\|^2 + OPT^2 = OPT^2 + \|\beta\|^2$$

We want $\|\beta\|^2 \leq 2\varepsilon OPT^2$. Since $\Pi A, \Pi U$ have same column span,

$$\Pi U(\alpha + \beta) = \Pi A\tilde{x}^* = \text{Proj}_{\Pi A}(\Pi b) = \text{Proj}_{\Pi U}(\Pi b)$$
$$= \text{Proj}_{\Pi U}(\Pi(U\alpha + w)) = \Pi U\alpha + \text{Proj}_{\Pi U}(\Pi w)$$

so $\Pi U\beta = \text{Proj}_{\Pi U}(\Pi w)$, so $(\Pi U)^T(\Pi U)\beta = (\Pi U)^T \Pi w$. Now, let Π be a $(1 - 1/\sqrt[4]{2})$-subspace embedding — then ΠU has smallest singular value at least $1/\sqrt[4]{2}$. Therefore

$$\|\beta\|^2/2 \leq \|(\Pi U)^T(\Pi U)\beta\|^2 = \|(\Pi U)^T \Pi w\|^2$$

Now suppose Π also approximately preserves matrix multiplication. Here w is orthogonal to the columns of A, so $U^T w = 0$. Then, by the general approximate matrix multiplication property,

$$\mathbb{P}_\Pi \left(\|(\Pi U)^T \Pi w - U^T w\|_2^2 > \varepsilon'^2 \|U\|_F^2 \|w\|_2^2 \right) < \delta$$

We have $\|U\|_F = \sqrt{d}$, so set error parameter $\varepsilon' = \sqrt{\varepsilon/d}$ to get

$$\mathbb{P} \left(\|(\Pi U)^T \Pi w\|^2 > \varepsilon \|w\|^2 \right) < \delta$$

so $\|\beta\|^2 \leq 2\varepsilon \|w\|^2 = 2\varepsilon OPT^2$, as required.

Ultimately, Π need not be an ε-subspace embedding. It suffices to merely be a c-subspace embedding for some fixed constant $c = 1 - 1/\sqrt{2}$, while giving approximate matrix multiplication with error $\sqrt{\varepsilon/d}$. Thus using the Thorup-Zhang sketch, this reduction we only require $m = O(d^2 + d/\varepsilon)$ and even $s = 1$, as opposed to the first reduction that needed $m = \Omega(d^2/\varepsilon^2)$.

3.6 Low-Rank Approximation

The basic idea is a matrix $A \in \mathbb{R}^{n \times d}$ with n, d both large, e.g. n users rating d movies. Suppose users are linear combinations of a few (k) basic types. We want to discover this low-rank structure.

Given a matrix $A \in \mathbb{R}^{n \times d}$, we want to compute $A_k := \text{argmin}_{\text{rank}(B) \leq k} \|A - B\|_X$.

Some now argue that we should look for a non-negative matrix factorization; nevertheless, this version is still used.

Theorem 3.5 (Eckart-Young) *Let* $A = U \Sigma V^T$ *be a singular-value decomposition of* A *where* $\text{rank}(A) = r$ *and* Σ *is diagonal with entries* $\sigma_1 \geq \sigma_2 \geq \cdots \geq \sigma_r > 0$, *then under* $\| \cdot \|_X = \| \cdot \|_F$, $A_k = U_k \Sigma_k V_k^T$ *is the minimizer where* U_k *and* V_k *are the first* k *columns of* U *and* V *and* $\Sigma_k = \text{diag}(\sigma_1, \ldots, \sigma_k)$.

Our output is then U_k, Σ_k, V_k. We can calculate A_k in $O(nd^2)$ time, by calculating the SVD of A.

Definition 3.8 $\text{Proj}_A B$ is the projection of the columns of B onto the colspace(A).

Definition 3.9 Let $A = U \Sigma V^T$ be a singular decomposition. $A^+ = V \Sigma^{-1} U^T$ is called *Moore-Penrose pseudoinverse* of A.

Now recall subspace embedding and approximate matrix multiplication to compute $\tilde{A}_k$ with rank at most k such that $\|A - \tilde{A}_k\|_F \leq (1 + \varepsilon)\|A - A_k\|_F$, following Sarlós' approach (Sarlós 2006). The first works which got some decent error (like $\varepsilon\|A\|_F$) was due to Papadimitriou (Papadimitriou et al. 2000) and Frieze, Kanna and Vempala (Frieze et al. 2004).

Theorem 3.6 *Define* $\tilde{A}_k = \text{Proj}_{A\Pi^T, k}(A)$. *As long as* $\Pi \in \mathbb{R}^{m \times n}$ *is an* $1/2$ *subspace embedding for a certain* k-*dimensional subspace* V_k *and satisfies approximate matrix multiplication with error* $\sqrt{\varepsilon/k}$, *then*

$$\|A - \tilde{A}_k\|_F \leq (1 + O(\varepsilon))\|A - A_k\|_F,$$

where $\text{Proj}_{V_k}(A)$ *is the best rank* k *approximation to* $\text{Proj}_V(A)$, *i.e., projecting the columns of* A *to* V.

Firstly, let us verify that this algorithm is fast, and that compute $\text{Proj}_{A\Pi^T, k}(A)$ fast. To satisfy the conditions in the above theorem, we know that $\Pi \in \mathbb{R}^{m \times d}$ can be chosen with $m = O(k/\varepsilon)$ e.g. using a random sign matrix or slightly larger m using a faster subspace embedding. We need to multiply $A\Pi^T$. We can use a fast subspace embedding to compute $A\Pi^T$ fast, then we can compute the SVD of $A\Pi^T = U'\Sigma'V'^T$ in $O(nm^2)$ time. Let $[\cdot]_k$ denote the best rank-k approximation under Frobenius norm. We wish to compute $[U'U'^T A]_k = U'[U'^T A]_k$. Computing $U'^T A$ takes $O(mnd)$ time, then computing the SVD of $U'^T A$ takes $O(dm^2)$ time. It is better than the $O(nd^2)$ time to compute the SVD of A, but we can do better if we approximate. By using the right combination of subspace embeddings, for constant ε the scheme described here can be made to take $O(nnz(A)) + \tilde{O}(ndk)$ time (where $\tilde{O}$ hides $\log n$ factors). We will do instead for $O(nnz(A)) + \tilde{O}(nk^2)$.

We want to compute $\tilde{A}_k = argmin_{X:rank(X) \leq k}\|U'X - A\|_F^2$. If X^+ is the argmin without the rank constraint, then the $argmin$ with the rank constraint is $[U'X^+]_k = U'[X^+]_k$, where $[\cdot]_k$ denotes the best rank-k approximation under Frobenius error.

Rather than find X^+, we use *approximate regression* to find an approximately optimal $\tilde{X}$. That is, we compute $\tilde{X} = argmin_X \|\Pi'U'X - \Pi'A\|_F^2$ where Π' is an α-subspace embedding for the column space of U' (see U' has rank m). Then output is $U'[\tilde{X}]_k$.

$$\left(\frac{1+\alpha}{1-\alpha}\right) \cdot \|U'X^+ - A\|_F^2 \geq \|U'\tilde{X} - A\|_F^2$$

$$= \|(U'X^+ - A) + U'(\tilde{X} - X^+)\|_F^2$$

$$= \|U'X^+ - A\|_F^2 + \|U'(\tilde{X} - X^+)\|_F^2$$

$$= \|U'X^+ - A\|_F^2 + \|\tilde{X} - X^+\|_F^2$$

and thus $\|\tilde{X} - X^+\|_F^2 \leq O(\alpha) \cdot \|U'X^+ - A\|_F^2$. The second equality above holds since the matrix U' preserves Frobenius norms, and the first equality since $U'X^+ - A$ has a column space orthogonal to the column space of U'. Next, suppose $f, \tilde{f}$ are two functions mapping the same domain to $\mathbb{R}$ such that $|f(x) - \tilde{f}(x)| \leq \eta$ for all x in the domain. Then $f(argmin_x \tilde{f}(x)) \leq \min_x f(x) + 2\eta$. Now, let the domain be the set of all rank-k matrices, and let $f(Q) = \|U'X^+ - Q\|_F$ and $\tilde{f}(Q) = \|U'\tilde{X} - Q\|_F$. Then $\eta = \|U'X^+ - U'\tilde{X}\|_F = \|X^+ - \tilde{X}\|_F$. Therefore

$$\|U'[\tilde{X}]_k - A\|_F^2 = \|U'[\tilde{X}]_k - U'X^+\|_F + \|(I - U'U'^T)A\|_F^2$$

$$\leq (\|U'[X^+]_k - U'X^+\|_F + 2 \cdot \|X^+ - \tilde{X}\|_F)^2 + \|(I - U'U'^T)A\|_F^2$$

$$\leq (\|U'[X^+]_k - U'X^+\|_F + O(\sqrt{\alpha}) \cdot \|U'X^+ - A\|_F)^2 + \|(I - U'U'^T)A\|_F^2$$

$$= (\|U'[X^+]_k - U'X^+\|_F + O(\sqrt{\alpha}) \cdot \|U'X^+ - A\|_F)^2 + \|U'X^+ - A\|_F^2$$

$$= \|U'[X^+]_k - U'X^+\|_F^2 + O(\sqrt{\alpha}) \cdot \|U'[X^+]_k - U'X^+\|_F \cdot \|U'X^+ - A\|_F$$

$$+ O(\alpha) \cdot \|U'X^+ - A\|_F^2 + \|U'X^+ - A\|_F^2$$

$$= \|U'[X^+]_k - A\|_F^2 + O(\sqrt{\alpha}) \cdot \|U'[X^+]_k - U'X^+\|_F \cdot \|U'X^+ - A\|_F$$

$$+ O(\alpha) \cdot \|U'X^+ - A\|_F^2 \tag{3.2}$$

$$\leq (1 + O(\alpha)) \cdot \|U'[X^+]_k - A\|_F^2 + O(\sqrt{\alpha}) \cdot \|U'[X^+]_k - U'X^+\|_F \cdot \|U'X^+ - A\|_F$$

$$\tag{3.3}$$

$$\leq (1 + O(\alpha)) \cdot \|U'[X^+]_k - A\|_F^2 + O(\sqrt{\alpha}) \cdot \|U'[X^+]_k - A\|_F^2 \tag{3.4}$$

$$= (1 + O(\sqrt{\alpha})) \cdot \|U'[X^+]_k - A\|_F^2$$

where (3.2) used that $\|U'[X^+]_k - U'X^+ + U'X^+ - A\|_F^2 = \|U'[X^+]_k - A\|_F^2 + \|U'[X^+]_k - U'X^+\|_F^2$ since $U'X^+ - A$ has columns orthogonal to the column space of U'. Also, (3.3) used that

$$\|U'X^+ - A\|_F \leq \|U'[X^+]_k - A\|_F,$$

since $U'X^+$ is the best Frobenius approximation to A in the column space of U'. Ultimately, (3.4) again used

$$\|U'X^+ - A\|_F \leq \|U'[X^+]_k - A\|_F,$$

and also used the triangle inequality

$$\|U'[X^+]_k - U'X^+\|_F \leq \|U'[X^+]_k - A\|_F + \|U'X^+ - A\|_F \leq 2 \cdot \|U'[X^+]_k - A\|_F.$$

So, we have established the following theorem that follows from the above calculations and Theorem 3.6.

Theorem 3.7 *Let $\Pi_1 \in \mathbb{R}^{m_1 \times n}$ be a 1/2 subspace embedding for a certain k-dimensional subspace V_k, and suppose Π_1 also satisfies approximate matrix multiplication with error $\sqrt{\varepsilon/k}$. Let $\Pi_2 \in \mathbb{R}^{m_2 \times n}$ be an α-subspace embedding for the column space of U', where $A\Pi_1^T = U'\Sigma'V'^T$ is the SVD (and hence U' has rank at most m_1). Let $\tilde{A}'_k = U'[\tilde{X}]_k$ where*

$$\tilde{X} = \underset{X}{argmin} \, \|\Pi_2 U'X - \Pi_2 A\|_F^2.$$

Then $\tilde{A}'_k$ has rank k and

$$\|A - \tilde{A}'_k\|_F \leq (1 + O(\varepsilon) + O(\sqrt{\alpha}))\|A - A_k\|_F.$$

In particular, the error is $(1 + O(\varepsilon))\|A - A_k\|_F$ for $\alpha = \varepsilon$.

Further, we show that $\text{Proj}_{A\Pi^T,k}(A)$ actually is a good rank-k approximation to A (i.e. we prove Theorem 3.6).

Proof We denote the first k columns of U and V as U_k and V_k and the remaining columns by $U_{\bar{k}}$ and $V_{\bar{k}}$. Let Y be the column span of $\text{Proj}_{A\Pi^T}(A_k)$ and the orthogonal projection operator onto Y as P. Then,

$$\|A - \text{Proj}_{A\Pi^T,k}(A)\|_F^2 \leq \|A - PA\|_F^2 = \|A_k - PA_k\|_F^2 + \|A_{\bar{k}} - PA_{\bar{k}}\|_F^2$$

Then we can bound the second term in that sum:

$$\|A_{\bar{k}}\| = \|(I - P)A_{\bar{k}}\|_F^2 \leq \|A_{\bar{k}}\|_F^1$$

Now we just need to show that $\|A_k - PA_k\|_F^2 \leq \varepsilon\|A_{\bar{k}}\|_F^2$:

$$\|A - PA\|_F^2 = \|(A_k - (A\Pi^T)(A\Pi^T)^+ A_k)\|_F^2 \leq \|A_k - (A\Pi^T)(A\Pi^T)^+ A_k\|_F^2$$

$$= \|A_k^T - A_k^T(\Pi A^T)^+(\Pi A^T)\|_F^2$$

$$= \sum_{i=1}^{n} \|A_k^{T(i)} - A_k^T(\Pi A^T)^+(\Pi A^T)^{(i)}\|_2^2$$

Here superscript (i) means the ith column. Now we have a bunch of different approximate regression problems which have the following form:

$$\min_x \|\Pi A_k^T x - \Pi(A^T)^{(i)}\|_2,$$

which has optimal value $\tilde{x}^* = (\Pi A_k^T)^+ (\Pi A^T)^{(i)}$. Consider the problem $\min_x \|\Pi A_k^T x - (A^T)^{(i)}\|_2$ as original regression problem. In this case optimal x^* gives $A_k^T x^* = \mathrm{Proj}_{A_k^T}((A^T)^{(i)}) = (A_k^T)^{(i)}$. Now we can use the analysis on the approximate least square from last week.

Here, we have a bunch of w_i, β_i, α_i with $S = A_k^T = V_k \Sigma_k U_k^T$ and $b_i = (A^T)^{(i)}$. Here, $\|w_i\|^2 = \|Sx^* - b\|^2 = \|(A_k^T)^{(i)} - (A^T)^{(i)}\|^2$. Hence $\sum_i \|w_i\|^2 = \|A - A_k\|_F^2$. Conversely, $\sum_i \|\beta_i\|^2 = \|A_k^T - A_k^T (\Pi A_k^T)^+ (\Pi A^T)\|_F^2$. Since $(\Pi V_k)^T (\Pi V_k) \beta_i = (\Pi V_k)^T \Pi w_i$, if all singular values of ΠV_k are at least $1/2^{1/4}$, we have

$$\frac{\sum_i \|\beta_i\|^2}{2} \le \sum_i \|(\Pi V_k)^T (\Pi V_k) \beta_i\|^2 = \sum_i \|(\Pi V_k)^T \Pi w_i\|^2 = \|(\Pi V_k)^T \Pi G\|_F^T$$

where G has w_i as ith column.

$(\Pi V_k)^T \Pi G$ exactly same as approximate matrix multiplication of V_k and G. Since columns of G and V_k are orthogonal, we have $V_k^T G = 0$, hence if Π is a sketch for approximate matrix multiplication of error $\varepsilon' = \sqrt{\varepsilon/k}$, then

$$\mathbb{P}_\Pi(\|(\Pi V_k)^T (\Pi G)\|_F^2 > \varepsilon \|G\|_F^2) < \delta$$

since $\|V_k\|_F^2 = k$. Clearly $\|G\|_F^2 = \sum_i \|w_i\|^2 = \|A - A_k\|_F^2$, hence proved.

3.7 Compressed Sensing

Compressed or compressive sensing developed from questions raised about the efficiency of the conventional signal processing pipeline for compression, coding and recovery of natural signals, including audio, still images and video.

Nowadays varied sensing devices such as mobile phones and biomedical sensors are indispensable. Individually operating sensors normally form correlated sensor networks in large scale. Therefore, these sensors generate continuous flows of big sensing data that pose key challenges: how to sense and transmit massive spatio-temporal data in efficient manner. Many conventional distributed sensing schemes process input signals in the sensing devices to reduce the burden of network transmission. However, these conventional schemes are not well suited for resource limited sensing devices because of excessive energy and resource consumption. Compressive sensing sheds light on this problem by shifting the complexity burden of encoding process to the decoder. Compressive sensing enables to compress large amounts of inputs signals without much energy consumption. Recent advances in Compressive sensing reduce this computational burden even further by random sampling, so that Compressive sensing schemes are successfully applied to large-scale sensor networks.

Moreover, one encounters the task of inferring quantities of interest from measured information in computer science. For example, in signal and image processing, one would like to reconstruct a signal from measured data. When the information acquisition process is linear, the problem reduces to solving a linear system of equations. A compressible signal is one which is sparse in some basis, but not necessarily the standard basis. Here an approximately sparse signal is a sum of a sparse vector with a low-weight vector.

Consider $x \in \mathbb{R}^n$. If x is a k sparse vector, we could represent it in a far more compressed manner. Thus, we define a measure of how "compressible" a vector is as a measure of how close it is to being k sparse.

Definition 3.10 Let $x_{head(k)}$ be the k elements of largest magnitude in x. Let $x_{tail(k)}$ be the rest of x.

Therefore, we call x compressible if $\|x_{tail(k)}\|$ is small.

The goal here is to approximately recover x from few linear measurements. Consider we have a matrix Πx such that each the ith row is equal to $\langle \alpha_i, x \rangle$ for some $\alpha_1, \ldots, \alpha_m \in \mathbb{R}^n$. We want to recover a $\tilde{x}$ from ΠX such that $\|x - \tilde{x}\|_p \leq C_{\varepsilon,p,q} \|x_{tail(k)}\|_q$, where $C_{\varepsilon,p,q}$ is some constant dependent on ε, p and q. Depending on the problem formulation, I may or may not get to choose this matrix Π.

There are many practical applications in which approximately sparse vectors appear. Pixelated images, for example, are usually approximately sparse in some basis U. For example, consider an n by n image $x \in \mathbb{R}^{n^2}$. then $x = Uy$ for some basis U, and y is approximately sparse. Thus we can get measurements from ΠUy.

Assume that n is a power of two. Then:

1. Break the image x into squares of size four pixels.
2. Initialize a new image, with four regions R_1, R_2, R_3, R_4.
3. Each block of four pixels, b, in x has a corresponding single pixel in each of R_{1b}, R_{2b}, R_{3b}, and R_{4b} based on its location. For each block of four b:

   - Let the b have pixel values p_1, p_2, p_3, and p_4.
   - $R_{1b} \leftarrow \frac{1}{4}(p_1 + p_2 + p_3 + p_4)$
   - $R_{2b} \leftarrow \frac{1}{4}(p_1 - p_2 + p_3 - p_4)$
   - $R_{3b} \leftarrow \frac{1}{4}(p_1 - p_2 - p_3 + p_4)$
   - $R_{4b} \leftarrow \frac{1}{4}(p_1 - p_2 + p_3 - p_4)$

4. Recurse on R_1, R_2, R_3, and R_4.

Normally, pixels are relatively constant in certain regions. So, the values in all regions except for the first are usually relatively small. If you view images after this transform, the upper left hand regions will often be closer to white, while the rest will be relatively sparse. A signal is called sparse if most of its components are zero. In empirical sense, many real-world signals are compressible that they are well approximated by sparse signals often after an appropriate change of basis. This describes why compression techniques such as JPEG, MPEG, or MP3 work extremely well in practice.

The basic approach to taking photo is to first take a high-resolution photo in the standard basis. That means, a light magnitude for each pixel and then to compress the picture later using software tool. Because photos are usually sparse in an appropriate basis. The compressed sensing approach asks, then why not just capture the image directly in a compressed form, i.e. in a representation where its sparsity shines through? For example, one can store random linear combinations of light intensities instead of the light intensities themselves. This idea leads to a reduction in the number of pixels needed to capture an image at a given resolution. Another application of compressed sensing is in Magnetic resonance imaging (MRI), where reducing the number of measurements decreases the time necessary for a scan.

Theorem 3.8 (Candès et al. 2006; Donoho 2006) *There exists a $\Pi \in \mathbb{R}^{m \times n}$ with $m = O(k lg(n/k))$ and a poly-time algorithm Alg s.t. if $\tilde{x} = Alg(\Pi x)$ then $\|x - \tilde{x}\|_2 \leq O(k^{-1/2}) \|x_{tail(k)}\|_1$*

If x is actually k-spares, $2k$ measurements are necessary and sufficient.

3.8 The Matrix Completion Problem

A partial matrix is a rectangular array in which some entries are specified, while the remaining unspecified entries are free to be chosen from an indicated set. A completion of a partial matrix is a particular choice of values for the unspecified entries resulting in a conventional matrix. In matrix completion, the positive definite completion problem has received the most attention, due to its role in several applications in probability and statistics, image enhancement, systems engineering, etc. and to its relation with other completion problems including spectral norm contractions and Euclidean distance matrices which is important for the molecular conformation problem in chemistry. In a typical matrix completion problem, description of circumstances is sought in which choices for the unspecified entries may be made from the same set so that the resulting ordinary matrix over that set is of a desired type. A matrix completion problem asks whether a given partial matrix has a completion of a desired type; for example, the positive definite completion problem asks which partial Hermitian matrices have a positive definite completion. The properties of matrix completion problems have been inherited permutation similarity, diagonal matrix multiplication and principal submatrices. Completion problems have proved to be a useful perspective to study fundamental matrix structure.

In a typical matrix completion problem, description of circumstances is sought in which choices for the unspecified entries may be made from the same set S so that the resulting ordinary matrix over S is of a desired type. In the vast majority of cases that have been of interest in matrix completion problem.

While the problem of rank aggregation is old, modern applications – such as those found in web-applications like Netflix and Amazon – pose new challenges. First, the data collected are usually cardinal measurements on the quality of each item, such as 1–5 stars, received from voters. Second, the voters are neither experts in the rating domain nor experts at producing useful ratings. These properties manifest themselves in a few ways, including skewed and indiscriminate voting behaviours.

A motivation for the matrix completion or Netflix problem comes from user ratings of some products which are put into a matrix M. The entries M_{ij} of the matrix correspond to the j'th user's rating of product i. We assume that there exists an ideal matrix that encodes the ratings of all the products by all the users. However, it is not possible to ask every user his opinion about every product. We are only given some ratings of some users and we want to recover the actual ideal matrix M from this limited data. So matrix completion is the following problem:

Problem: Suppose you are given some matrix $M \in \mathbb{R}^{n_1 \times n_2}$. Moreover, you also are given some entries $(M_{ij})_{ij \in \Omega}$ with $|\Omega| \ll n_1 n_2$.

Goal: We want to recover the missing elements in M.

This problem is hard if we do not make any additional premises on the matrix M since the missing M_{ij} could in principle be arbitrary. We will consider a recovery scheme that relies on the following three premises.

1. M is (approximately) low rank.
2. Both the columns space and the row space are "incoherent". We say a space is incoherent, when the projection of any vector onto this space has a small ℓ_2 norm.
3. If $M = U \Sigma V^T$ then all the entries of $U V^T$ are bounded.
4. The subset Ω is chosen uniformly at random.

Note 3.1 There is work on adversarial recovery where the values are not randomly chosen but rather carefully picked to trick us by an adversary.

Under these premises we show that there exists an algorithm that needs a number of entries in M bounded by $|\Omega| \leq (n_1 + n_2) \, r \, \text{poly} \, (\log(n_1 n_2)) \cdot \mu$. Here μ captures to what extent properties 2 and 3 above hold. One would naturally consider the following recovery method for the matrix M:

$$\begin{aligned} \text{minimize} \quad & \text{rank}(X) \\ \text{subject to:} \quad & X_{ij} = M_{ij} \; \forall i, j \in \Omega. \end{aligned}$$

Alas, this optimization problem is NP-hard. Hence, let us consider the following alternative optimization problem in trace norm, or *nuclear norm*.

$$\begin{aligned} \text{minimize} \quad & \|X\|_* \\ \text{subject to:} \quad & X_{ij} = M_{ij} \; \forall i, j \in \Omega, \end{aligned}$$

where the nuclear norm of X defined as the sum of the singular values of X, i.e. $\|X\|_* = \sum_i \sigma_i(X)$. This problem is a semi-definite program (SDP), and can be solved in time polynomial in $n_1 n_2$.

While a several heuristics have been developed across many disciplines, the general problem of finding the lowest rank matrix satisfying equality constraints is NP-hard. Most low-rank matrices could be recovered from most sufficiently large sets of entries by computing the matrix of minimum nuclear norm that agreed with the provided entries, and moreover the revealed set of entries could comprise a vanishing fraction of the entire matrix. The nuclear norm is equal to the sum of the singular values of a matrix and is the best convex lower bound of the rank function on the set of matrices whose singular values are all bounded by 1. The intuition behind this heuristic is that whereas the rank function counts the number of non-vanishing singular values, the nuclear norm sums their amplitude. Moreover, the nuclear norm can be minimized subject to equality constraints via semi-definite programming.

3.8.1 Alternating Minimization

Alternating minimization is a widely used heuristic for matrix completion in which the goal is to recover an unknown low-rank matrix from a subsample of its entries. Alternating minimization has been used in the context of matrix completion and continues to play an important role in practical approaches to the problem. The approach also formed an important component in the winning submission for the Netflix Prize. The iterative procedure behind Alternating Minimization (AM) is given below. We try to find an approximate rank-k factorization $M \approx X \cdot Y$, where X has k columns and Y has k rows. We start off with initial X_0, Y_0. Then we do as follows:

1. initialize X_0, Y_0
2. **for** $\ell = 1, \ldots, T$:

   a. $X_\ell \leftarrow argmin_X \|R(M - XY_{\ell-1})\|_F^2$
   b. $Y_\ell \leftarrow argmin_Y \|R(M - X_\ell Y)\|_F^2$

3. **return** X_T, Y_T

Rigorous analyses of modifications of the above AM template have been carried out in (Hardt 2014; Hardt and Wootters 2014). The work (Schramm and Weitz 2015) has also shown some performance guarantees when the revealed entries are *adversarial* except for random.

Now let us elaborate the main theorem and related definitions.

Definition 3.11 Let $M = U\Sigma V^*$ be the singular value decomposition. (See that $U \in \mathbb{R}^{n_1 \times r}$ and $V \in \mathbb{R}^{n_2 \times r}$.)

Definition 3.12 Define the incoherence of the subspace U as $\mu(U) = \frac{n_1}{r} \cdot \max_i \|P_U e_i\|^2$, where P_U is projection onto U. Similarly, the incoherence of V is $\mu(V) = \frac{n_2}{r} \cdot \max_i \|m P_V e_i\|^2$, where P_V is projection onto V.

Definition 3.13 $\mu_0 \overset{\text{def}}{=} \max\{\mu(U), \mu(V)\}$.

Definition 3.14 $\mu_1 \overset{\text{def}}{=} \|UV^*\|_\infty \sqrt{n_1 n_2 / r}$, where $\|UV\|_\infty$ is the largest magnitude of an entry of UV.

Theorem 3.9 *If $m \gtrsim \max\{\mu_1^2, \mu_0\} \cdot n_2 r \log^2(n_2)$ then with high probability M is the unique solution to the semi-definite program* $\min \|X\|_*$ *s.t.* $\forall i, j \in \Omega, X_{ij} = M_{ij}$.

We know that $1 \leq \mu_0 \leq \frac{n_2}{r}$.

μ_0 can be $\frac{n_2}{r}$ since a standard basis vector appears in a column of V, and μ_0 can get down to 1 is a kind of best case scenario where all the entries of V are similar to $\frac{1}{\sqrt{n_2}}$. Further, all the entries of U are similar to $\frac{1}{\sqrt{n_1}}$, if you took a Fourier matrix and remove some of its columns. Ultimately, the condition on m is a good bound if the matrix has low incoherence.

Reference (Candès and Tao 2010) proved that $m \gtrsim \mu_0 n_2 r \log(n_2)$ is essential. If you want to recover M over the random choice of Ω via SDP, then you need to sample at least that many entries. The condition isn't entirely compact because of the square in the log factor and the dependence on μ_1^2. However, Cauchy-Schwarz inequality implies $\mu_1^2 \leq \mu_0^2 r$.

The algorithm looks as follows when we want to minimize $\|AX - M\|_F^2 + \mu\|X\|_*$:

Select X_0, and a stepsize t and iterate (a)–(d) some number of times:

(a) $W = X_k - t \cdot A^T(AX_k - M)$
(b) $[U, \mathbf{diag}(s), V] = \mathbf{svd}(Q)$
(c) $r = \max(s - \mu t, 0)$
(d) $X_{k+1} = U\mathbf{diag}(r)V^T$

Definition 3.15 $\langle A, B \rangle \stackrel{\text{def}}{=} Tr(A^*B) = \sum_{i,j} A_{ij} B_{ij}$

Claim *The dual of the trace norm is the operator norm:*

$$\|A\|_* = \sup_{\substack{B \ s.t. \\ \|B\| \leq 1}} \langle A, B \rangle$$

This is logical since the dual of ℓ_1 for vectors is ℓ_∞. Furthermore, the trace norm and operator norm are similar to the ℓ_1 and ℓ_∞ norm of the singular value vector respectively.

Lemma 3.1

$$\underbrace{\|A\|_*}_{(1)} = \underbrace{\min_{\substack{X,Y \ s.t. \\ A=XY^*}} \|X\|_F \cdot \|Y\|_F}_{(2)} = \underbrace{\min_{\substack{X,Y \ s.t. \\ A=XY^*}} \frac{1}{2}\left(\|X\|_F^2 + \|Y\|_F^2\right)}_{(3)}$$

Proof **(2) $\leq$ (3):**
  AM-GM inequality: $xy \leq \frac{1}{2}(x^2 + y^2)$.
(3) $\leq$ (1):
  We simply need to show an X and Y which gives $\|A\|_*$.
  Set $X = Y^* = A^{1/2}$. Given $f : \mathbb{R}^+ \mapsto \mathbb{R}^+$, then $f(A) = Uf(\Sigma)V^*$. i.e. write the SVD of A and apply f to each diagonal entry of Σ. It is simple to verify that

$A^{1/2}A^{1/2} = A$ and that the square of the Frobenius norm of $A^{1/2}$ is exactly the trace norm.

(1) ≤ (2):

Let X, Y be some matrices such that $A = XY^*$. Then

$$\|A\|_* = \|XY^*\|_*$$

$$\leq \sup_{\substack{\{a_i\} \text{ orthonormal basis} \\ \{b_i\} \text{ orthonormal basis}}} \sum_i \langle XY^*a_i, b_i \rangle$$

$$= \sup_{\cdots} \sum_i \langle Y^*a_i, X^*b_i \rangle$$

$$\leq \sup_{\cdots} \sum_i \|Y^*a_i\| \cdot \|X^*b_i\|$$

$$\leq \sup_{\cdots} \left(\sum_i \|Y^*a_i\|^2 \right)^{1/2} \left(\sum_i \|X^*b_i\|^2 \right)^{1/2} \tag{3.5}$$

$$= \|X\|_F \cdot \|Y\|_F$$

Proof $\|A\|_* \leq \sup_{\|B\|=1} \langle A, B \rangle$.

By taking $A = U\Sigma V^*$ and $B = \sum_i u_i v_i^*$ we get the trace norm.

$$\|A_*\| \geq \langle A, B \rangle \quad \forall B \text{ s.t. } \|B\| = 1.$$

- Write $A = XY^*$ s.t. $\|A\|_* = \|X\|_F \cdot \|Y\|_F$.
- Write $B = \sum_i \sigma_i a_i b_i, \forall i, \sigma_i \leq 1$.

Then using a similar argument to (3.5),

$$\langle A, B \rangle = \langle XY^*, \sum_i \sigma_i a_i b_i \rangle$$

$$= \sum_i \sigma_i \langle Y^*a_i, X^*b_i \rangle$$

$$\leq \sum_i |\langle Y^*a_i, X^*b_i \rangle|$$

$$\leq \|X\|_F \|Y\|_F = \|A\|_*$$

hence proved.

While the principle of alternating maximization is well known in the literature, it had not been used before in the context of the present topic. Since it is not a matrix decomposition method, it can also be adapted to large-scale problems using essentially actions of matrix exponentials on vectors.

Chapter 4
Assorted Computational Models

This chapter presents some other computational models to tackle massive datasets efficiently. We will see formalized models for some massive data settings, and explore core algorithmic ideas arising in them. The models discussed are cell probe, online bipartite matching, MapReduce programming model, Markov chain, and crowd-sourcing. Finally, we present some basic aspects of communication complexity.

4.1 Cell Probe Model

The cell-probe model is one of the significant models of computation for data structures, subsuming in particular the common word-Random-access machine (RAM) model. We suppose that the memory is divided into fixed-size cells (words), and the cost of an operation is just the number of cells it reads or writes. Let $U = [m]$ be a universe of size m, and let $S \subseteq U$ with $|S| = n$. An algorithm is supplied with S, and it provides answer queries on elements of S or even U. The set S is kept in memory in *cells*, each of $\log m$ bits. The algorithm is executed in the following two stages:

1. Preprocessing: On receiving S, store S in memory in some suitable form. We denote the space utilised, measured in number of cells, by s.
2. Query: Given $x \in U$, return some information about x depending on the problem. Let t denote the maximum number of memory cell probes the algorithm takes to process each query.

The performance of the algorithm is measured only by the parameters s and t. The time taken by the preprocessing step is not counted. In the 'Query' step, number of memory probes is significant. Further, no information is carried over from the preprocessing to the query stage unless explicitly stored in the data structure. One can imagine this as two distinct algorithms: one for preprocessing and one for queries.

For each S, once it has been preprocessed, the execution of the Query algorithm can be represented by a decision tree. Given an $x \in U$, the algorithm proceeds as follows: depending on x it chooses a memory cell and *probes* it (reads its contents).

R. Akerkar, *Models of Computation for Big Data*, SpringerBriefs in Advanced Information and Knowledge Processing, https://doi.org/10.1007/978-3-319-91851-8_4

Depending on the contents of that cell it probes some other cell, and so on. Let us fix S. For every $x \in U$, we have a decision tree. The vertices are labeled by pointers to memory cells. The root's label is the pointer to the cell probed first. Each vertex has a child for every possible outcome of probing the location it points to. If t is the query complexity, then the depth of each tree is at most t.

4.1.1 The Dictionary Problem

In the dictionary problem, we are given an S which we need to store in memory in some form. For each $x \in S$, we will have a memory cell in a data structure T, which contains x and a pointer to some memory location containing auxiliary data about x. Furthermore, the algorithm might allocate some additional cells which will help it in processing queries.

Given $u \in U$, the problem is to find i such that $T[i] = u$, or report that $u \notin S$. The aim is to simultaneously reduce s and t.

A completely simple approach is to store the characteristic bit vector of the set S in the preprocessing phase, and answer every query with a single probe. This scheme has $s = m$ and $t = 1$.

The standard approach is to maintain a sorted array for storing S, and to use binary search for locating elements. For this scheme, $s = O(n)$ and $t = O(\log n)$.

For the dictionary problem we will use the *Fredman–Komlós–Szemerédi (FKS) scheme* from (Fredman et al. 1984). It achieves $s = O(n)$ and $t = O(1)$.

Theorem 4.1 (Fredman–Komlós–Szemerédi) *There exists a solution to the dictionary problem with $s = O(n)$ and $t = O(1)$.*

In the preprocessing phase, a good hashing function $h : [m] \longrightarrow [n]$ maps S without collisions. The algorithm would then store a description of h, and information about $x \in S$ in the cell numbered $h(x)$. In the Query phase, the algorithm on input x would read h, compute $h(x)$ and look up that cell. But h must have a compact description, otherwise reading h itself will need too many probes. We can find an h with a compact description which, though not collision-free, results in sufficiently small buckets. Then for each bucket, we can find a second-level hash function that is collision-free and has a compact description. Putting this together, both the storage requirement and the number of probes will be small.

Let us begin with a claim. Select a hash function $h : [m] \longrightarrow [n]$ uniformly at random from a family $\mathscr{H}$ of pairwise independent hash functions. For $i \in [n]$, let S_i be the ith bucket; $S_i = \{j \in S : h(j) = i\} = h^{-1}(i) \cap S$. Let K_i be the size of the ith bucket; $K_i = |h^{-1}(i) \cap S|$. The claim below shows that the expected sum of the squared bucket sizes is $Ox(n)$.

Claim $E_h\left[\sum_{i\in[n]} K_i^2\right] = O(n)$.

Proof $\forall u, v \in S$, let $\chi_{u,v}$ be the indicator variable of the event $h(u) = h(v)$ over the choice of h. Now, for each $u \in S$, $\Sigma_{v\in S}\chi_{u,v}$ is the number of elements in S that u clashes with, and is hence equal to $K_{h(u)}$. Since for all $u \in S_i$ the sum $\Sigma_{v\in S}\chi_{u,v}$ is equal to K_i, and since $|S_i| = K_i$, we have $E[\Sigma_{i\in[n]}K_i^2] = E[\Sigma_{i\in[n]}\Sigma_{u\in S_i}\Sigma_{v\in S}\chi_{u,v}] = E[\Sigma_{u,v\in S}\chi_{u,v}] = n + n(n-1)/n \leq 2n$.

Now we can present the preprocessing algorithm. For every $S \subseteq [m]$ of size n, aforementioned claim guarantees the existence of a hash function $h : [m] \longrightarrow [n]$ for which $E[\Sigma_{i\in[n]}K_i^2] = O(n)$. Fix such an h. From now onwards we will use K_i to denote the value taken by the random variable K_i when this h that we have fixed is chosen as the hash function. For each i with $K_i \neq 0$, let $\mathscr{H}_i$ be a family of pairwise independent hash functions $[m] \longrightarrow [2K_i^2]$. If we pick an h_i randomly from this family, then the probability that h_i has a collision within S_i is at most $\frac{1}{2K_i^2}\binom{K_i}{2} \leq 1/4$. Thus there is one function h_i which is collision-free within S_i. For each i fix one such h_i. The algorithm proceeds as follows. Given S, it determines $h, h_1, \ldots, h_n$ as above. It allocates n chunks of memory, the ith being of size $2K_i^2$. Call the ith chunk C_i. An array of n cells is allocated, one cell for each h_i. The ith cell, say p_i, contains the address of the first cell of C_i. An additional array of size $n + 1$ is used to store a description of the functions $h, h_1, \ldots, h_n$. The storage required is thus $O(n)$ for all the chunks together, plus whatever is required to store the functions.

The Query algorithm on input $x \in [m]$ proceeds as follows: Read the description of h and compute $h(x) = i$. Read cell p_i and the description of h_i. Adding the contents of cell p_i to $h_i(x)$ gives a location $m(x)$. If the cell at this location does not contain x, then conclude that $x \notin S$. If it does, then read on for auxillary information about x. The choices of h and h_i's ensure that for every $x \in S$ we are mapped to a distinct memory cell.

We need now to describe how we efficiently store and compute the functions h and h_i's. Take $\mathscr{H} = \{(ax + b) \bmod n : a, b \in [m], a \neq 0\}$ and $\mathscr{H}_i = \{(ax + b) \bmod 2K_i^2 : a, b \in [m], a \neq 0\}$. Then h and each h_i can be described using 2 cells (for storing a and b) and computable in constant time. Hence $s \in O(n)$ and $t \in O(1)$. Hence proved.

4.1.2 The Predecessor Problem

For a non-empty set S, and for every $x \in U$, the predecessor $\mathsf{Pred}_S(x)$ is defined as

$$\mathsf{Pred}(x) = \begin{cases} \max\{y \in S : y \leq x\} & \text{if such a } y \in S \text{ exists} \\ -1 & \text{otherwise} \end{cases}$$

We will prove an upper bound on s and t for the Predecessor Problem. We will present an algorithm for which $s = O(n\log m)$ and $t = O(\log\log m)$, which is not the best procedure. The efficient algorithm for the cell probe model is due to *Beame and Fich* (Beame and Fich 2002), who show how to achieve $s = O(n \log m)$ and $t = O(\min\{\frac{\log \log m}{\log \log \log m}, \sqrt{\frac{\log n}{\log \log n}}\})$. They have shown this bound to be tight for deterministic algorithms.

We use *X-tries* which is known as van Emde Boas trees, see (Cormen et al. 2009) to design a solution. One can think of each element of $U = [m]$ as a $\log m$ bit binary string. We build a complete binary tree of depth $\log m$, whose leaves correspond to elements of $[m]$. Each edge from a vertex to its left child is labelled 0 and each edge from a vertex to its right child is labelled 1, and the labels of the edges along the path from the root to a leaf u when concatenated give the binary representation of u. Call this tree T. We edit this tree by deleting all leaves that correspond to vertices not in S, all vertices that become leaves because of these deletions and so on. Finally we have a binary tree whose leaves are exactly the elements of S. Call it T'. The number of leaves in T' is n. As in every intermediate level there can be at most n vertices, and there are $O(\log m)$ levels, the size of T' is $O(n \log m)$. See that every element of S is its own predecessor. For an element $u \in U \setminus S$, let v be the deepest ancestor of u in T that is also present in T'. (Such a v must exist since at least one ancestor of u, the root of T, is in T'.) By the construction above, v is not a leaf in T'. Now there are two possibilities.

1. u is in the right subtree of v in T. By choice of v, v does not have a right child in T', but is not a leaf, so it has a left child. Clearly in this case the predecessor of u is the right most leaf of the left subtree of v in T'. In the preprocessing step we will identify such vertices v and create a link pointing from v to the rightmost leaf of its left subtree.
2. u is in the left subtree of v in T. By choice of v, v does not have a left child in T'. To find a predecessor, we need to go up from v until we find a vertex with a left child, go to the left subtree, and report the rightmost leaf there. So in the proprocessing step we put a link from v to the rightmost leaf in the left subtree rooted at the deepest ancestor of v with a left child. If there is no such ancestor of v in T', then we link from v to a special vertex that will denote a -1 value for Pred_S.

Therefore for each $u \in U$, once we get to the vertex v, we immediately obtain the predecessor by following the links. Thus we will be able to find the vertex v, given u. Let the bit string corresponding to u be $b_1 b_2 \ldots b_k$ where $k = O(\log m)$. So the path from the root to the vertex v we are looking for is $b_1 \ldots b_p$ where $p = \max\{x \le k : b_1 \ldots b_x$ is the label of a vertex in $T'\}$. (Note: $x = k$ exactly when $u \in S$.) The idea is to do a binary search in T' to identify v. If we can check whether a binary string $b_1 b_2 \ldots b_l$ forms a path from the root to some vertex in T' with $O(1)$ probes, then we can get to the vertex v with $O(\log k)$ probes. This will give us the essential $O(\log \log m)$ probe result.

4.2 Online Bipartite Matching

Introduced in 1990 by Karp, Vazirani, and Vazirani (Karp et al. 1990), on-line bipartite matching was one of the first problems to receive the attention of competitive analysis. In recent years, the problem of maximum online bipartite matching with dynamic posted prices, motivated by the real-world challenge of efficient parking allocation. Smart parking systems are being deployed in an increasing number of cities. Such systems allow commuters and visitors to see in real time, using cellphone applications or other digital methods, all available parking slots and their prices. In the original bipartite matching problem we seek to find a maximum matching, i.e. a matching that contains the largest possible number of edges given a graph.

On the other hand, in a "online" bipartite matching problem, we observe vertices one by one and assign matchings in an online fashion. Our goal is to find an algorithm that maximizes the competitive ratio $R(A)$.

Definition 4.1 (*Competitive ratio*)

$$R(A) := \liminf_I \frac{e[\mu_A(I)]}{\mu_*(I)} \tag{4.1}$$

where $\mu_A(I)$ and $\mu_A(I)$ denote the size of matching for an algorithm A and maximum matching size respectively, given input $I := \{\text{graph, arriving order}\}$.

Obviously $R(A) \leq 1$, but can we find a lower bound for $R(A)$?

4.2.1 Basic Approach

Since each edge can block at most two edges, we have $R(A) \geq 0.5$. On the other hand, for any deterministic algorithm A, we can find an adversarial input I such that $R(A) \leq 0.5$.

Consider the graph, where there is a perfect matching from n vertices on the left to n vertices on right, and the second half of us are fully connected to the first half of v. Under this setting, the number of correctly matched vertices in the second half of v is at most $n/2$. The expected number of correctly matched vertices in the first half is given by:

$$e[\text{\#correctly matched vertices}] = \sum_{i=1}^{n/2} \mathbb{P}[i\text{th vertex is correctly matched}] \tag{4.2}$$

$$\leq \sum_{i=1}^{n/2} \frac{1}{\frac{n}{2} - i + 2} \tag{4.3}$$

$$\leq \log\left(\frac{n}{2} + 1\right) \tag{4.4}$$

Since $\mu_* = n$, the competitive ratio R:

$$R(A) = \frac{e[\#matched]}{n} \leq \frac{\frac{n}{2} + log(\frac{n}{2} + 1)}{n} \rightarrow \frac{1}{2}$$

This randomized algorithm does not do better than $1/2$.

4.2.2 Ranking Method

Consider a graph G with appearing order π. Without selecting a random edge, we randomly permute the v's with permutation $\sigma(\cdot)$. We then match u to

$$v := \underset{v' \in \mathcal{N}(u)}{\arg\min} \sigma(v')$$

where $\mathcal{N}(u)$ denotes the neighbors of u.

Let us prove that this algorithm achieves a competitive ratio of $1 - 1/e$. We begin by defining our notation. The matching is denoted by Matching(G, π, σ). $M^*(v)$ denotes the vertex matched to v in perfect matching. $G := \{U, V, E\}$, where U, V, E denote left vertices, right vertices and edges respectively.

Lemma 4.1 *Let $H := G - \{x\}$ with permutation π_H and arriving order σ_H induced by π, σ respectively. Matching$(H, \pi_H, \sigma_H) =$ Matching$(G, \pi, \sigma) +$ augmenting path from x downwards.*

Lemma 4.2 *Let $u \in U$ and $M^*(u) = v$, if v is not matched under σ, then u is matched to v' with $\sigma(v') \leq \sigma(v)$.*

Lemma 4.3 *Let x_t be the probability that the rank-t vertex is matched. Then*

$$1 - x_t \leq \frac{\sum_{s \leq t} x_s}{n} \tag{4.5}$$

Proof Let v be the vertex with $\sigma(v) = t$. Note, since σ is uniformly random, v is uniformly random. Let $u := M^*(v)$. Denote by R_t the set of left vertices that are matched to rank $1, 2, \ldots, t$ vertices on the right. We have $e[|R_{t-1}|] = \sum_{s \leq t-1} x_s$. If v is not matched, u is matched to some $\tilde{v}$ such that $\sigma(\tilde{v}) < \sigma(v) = t$, or equivalently, $u \in R_{t-1}$. Hence,

$$\mathbb{P}(v \text{ not matched}) = 1 - x_t = \mathbb{P}(u \in R_{t-1}) = \mathbb{P}\left(\frac{e[|R_{t-1}|]}{n}\right) \leq \frac{\sum_{s \leq t} x_s}{n}$$

However this proof is not correct since u and R_{t-1} are not independent and thus $\mathbb{P}(u \in R_{t-1}) \neq \mathbb{P}(\frac{e[|R_{t-1}|]}{n})$. Instead, we use the following lemma to complete the correct proof.

Lemma 4.4 *Given σ, let $\sigma^{(i)}$ be the permutation that is σ with v moved to the ith rank. Let $u := M^*(v)$. If v is not matched by σ, for every i, u is matched by $\sigma^{(i)}$ to some $\tilde{v}$ such that $\sigma^{(i)}(\tilde{v}) \leq t$.*

Proof By Lemma 4.1, inserting v to ith rank causes any change to be a move up.

$$\sigma^{(i)}(\tilde{v}) \leq \sigma(\tilde{v}) + 1 \leq t$$

Proof (By Lemma 4.3) Given σ, let $\sigma^{(i)}$ be the permutation that is σ with v moved to the ith rank, where v is picked uniformly at random. Let $u := M^*(v)$. If v is not matched by σ (with probability $1 - x_t$), then u is matched by σ' to some $\tilde{v}$ such that $\sigma(\tilde{v}) \leq t$, or equivalently $u \in R_t$.

Choose random σ and v, let $\sigma' = \sigma$ with v moved to rank t. $u := M^*(v)$. According to Lemma 4.4, if v is not matched by σ (with probability x_t), u in σ' is matched to $\tilde{v}$ with $\sigma'(\tilde{v}) \leq t$, or equivalently $u \in R_t$. Note, u and R_t are now independent and $\mathbb{P}(u \in R_t) = |R_t|/n$ holds. Hence proved.

With Lemma 4.3, we can finally obtain the final results. Let $s_t := \sum_{s \leq t} x_s$. Lemma 4.3 is equivalent to $s_t(1 + 1/n) \geq 1 + s_{t-1}$. Solving the recursion, it can also be rewritten as $s_t = \sum_{s \leq t}(1 - 1/(1 + n))^s$ for all t. The competitive ratio is thus, $s_n/n \to 1 - 1/e$.

4.3 MapReduce Programming Model

A growing number of commercial and science applications in both classical and new fields process very large data volumes. Dealing with such volumes requires processing in parallel, often on systems that offer high compute power. Such type of parallel processing, the MapReduce paradigm (Dean and Ghemawat 2004) has found popularity. The key insight of MapReduce is that many processing problems can be structured into one or a sequence of phases, where a first step (Map) operates in fully parallel mode on the input data; a second step (Reduce) combines the resulting data in some manner, often by applying a form of reduction operation. MapReduce programming models allow the user to specify these map and reduce steps as distinct functions; the system then provides the workflow infrastructure, feeding input data to the map, reorganizing the map results, and then feeding them to the appropriate reduce functions, finally generating the output.

While data streams are an efficient model of computation for a single machine, MapReduce has become a popular method for large-scale parallel processing.

In MapReduce model, data items are each $\langle key, value \rangle$ pairs. For example, you have a text file 'input.txt' with 100 lines of text in it, and you want to find out the frequency of occurrence of each word in the file. Each line in the input.txt file is considered as a value and the offset of the line from the start of the file is considered as a key, here (offset, line) is an input $\langle key, value \rangle$ pair. For counting how many times a word occurred (frequency of word) in the input.txt, a single word is considered as an output key and a frequency of a word is considered as an output value.

Our input ⟨*key*, *value*⟩ is (offset of a line, line) and output ⟨*key*, *value*⟩ is (word, frequency of word).

A Map-Reduce job is divided into four simple phases, Map phase, Combine phase, Shuffle phase, and Reduce phase:

- *Map*: Map function operates on a single record at a time. Each item is processed by some *map* function, and emits a set of new ⟨*key*, *value*⟩ pairs.
- *Combine*: The combiner is the process of applying a reducer logic early on an output from a single map process. Mappers output is collected into an in memory buffer. MapReduce framework sorts this buffer and executes the commoner on it, if you have provided one. Combiner output is written to the disk.
- *Shuffle*: In the shuffle phase, MapReduce partitions data and sends it to a reducer. Each mapper sends a partition to each reducer. This step is natural to the programmer. All items emitted in the map phase are grouped by key, and items with the same key are sent to the same reducer.
- *Reducer*: During initialization of the reduce phase, each reducer copies its input partition from the output of each mapper. After copying all parts, the reducer first merges these parts and sorts all input records by key. In the Reduce phase, a reduce function is executed only once for each key found in the sorted output. MapReduce framework collects all the values of a key and creates a list of values. The Reduce function is executed on this list of values and a corresponding key. So, Reducer receives ⟨$k, v_1, v_2, \ldots, v_3$⟩ and emits new set of items.

MapReduce provides many significant advantages over parallel databases. Firstly, it provides fine-grain fault tolerance for large jobs; failure in the middle of a multi-hour execution does not require restarting the job from scratch. Secondly, MapReduce is very useful for handling data processing and data loading in a heterogeneous system with many different storage systems. Third, MapReduce provides a good framework for the execution of more complicated functions than are supported directly in SQL.

Data streaming and MapReduce have emerged as two leading paradigms for handling computation on very large datasets. As the datasets have grown to tera- and petabyte input sizes, two paradigms have emerged for developing algorithms that scale to such large inputs: streaming and MapReduce (Bahmani et al. 2012). In the streaming model, as we have seen, one assumes that the input can be read sequentially in a number of passes over the data, while the total amount of random access memory (RAM) available to the computation is sublinear in the size of the input. The goal is to reduce the number of passes needed, all the while minimizing the amount of RAM necessary to store intermediate results. In the case the input is a graph, the vertices V are known in advance, and the edges are streamed. The challenge in streaming algorithms lies in wisely using the limited amount of information that can be stored between passes.

Complementing streaming algorithms, MapReduce, and its open source implementation, Hadoop, has become the *de facto* model for distributed computation on a massive scale. Unlike streaming, where a single machine eventually sees the whole dataset, in MapReduce, the input is partitioned across a set of machines, each of which can perform a series of computations on its local slice of the data. The process can then be repeated, yielding a multi-pass algorithm. It is well known that simple operations like sum and other holistic measures as well as some graph primitives, like finding connected components, can be implemented in MapReduce in a work-efficient manner. The challenge lies in reducing the total number of passes with no machine ever seeing the entire dataset.

4.4 Markov Chain Model

Randomization can be a useful tool for developing simple and efficient algorithms. So far, most of these algorithms have used independent coin tosses to generate randomness. In 1907, A. A. Markov began the study of an important new type of chance process. In this process, the outcome of a given experiment can affect the outcome of the next experiment. This type of process is called a Markov chain (Motwani and Raghavan 1995). Specifically, Markov Chains represent and model the flow of information in a graph, they give insight into how a graph is connected, and which vertices are important.

A *random walk* is a process for traversing a graph where at every step we follow an outgoing edge chosen uniformly at random. A *Markov chain* is similar except the outgoing edge is chosen according to an arbitrary fixed distribution.

One use of random walks and Markov chains is to sample from a distribution over a large universe. In general, we set up a graph over the universe such that if we perform a long random walk over the graph, the distribution of our position approaches the distribution we want to sample from. Given a random walk or a Markov chain we would like to know: How quickly can we reach a particular vertex; How quickly can we cover the entire graph? How quickly does our position in the graph become "random"? While random walks and Markov chains are useful algorithmic techniques, they are also useful in analyzing some natural processes.

Definition 4.2 (*Markov Chain*) A Markov Chain $(X_t)_{t \in \mathbb{N}}$ is a sequence of random variables on some state space S which obeys the following property:

$$\forall t > 0, (s_i)_{i=0}^{t} \in S, \mathbb{P}\left[X_t = s_t \,\middle|\, \bigcap_{i=0}^{t-1}(X_i = s_i)\right] = \mathbb{P}[X_1 = s_t | X_0 = s_{t-1}]$$

We take these probabilities as a *transition matrix* P, where $P_{ij} = \mathbb{P}[X_1 = s_j | X_0 = s_i]$. See that $\forall i, \sum_j P_{ij} = 1$ is necessary for P to be a valid transition matrix.

If $q \in \mathbb{R}^{|S|}$ is the distribution of X at time 0, the distribution of X at time t will then be qP^t.

Theorem 4.2 (The Fundamental Theorem of Markov Chains) *Let X be a Markov Chain on a finite state space $S = [n]$ satisfying the following conditions:*

Irreducibility   There is a path between any two states which will be followed with > 0 probability, i.e. $\forall i, j \in [n], \exists t \mathbb{P}[X_t = j | X_0 = i] > 0$.

Aperiodicity   Let the period *of a pair of states u, v be the GCD of the length of all paths between them in the Markov chain, i.e.* $\gcd\{t \in \mathbb{N}_{>0} | \mathbb{P}[X_t = v | X_0 = u] > 0\}$. *X is aperiodic if this is 1 for all u, v.*

Then *X is ergodic.*

These conditions are necessary as well as sufficient.

$$N(i, t) = |\{t \in \mathbb{N} | X_t = i\}|$$

This follows $\lim_{t \to \infty} \frac{N(i,t)}{t} = \Pi_i$ for an ergodic chain with stationary distribution Π.

$$h_{u,v} = \mathbb{E}[\min_t \{t | X_t = v\} | X_0 = u]$$

This is called the *hitting time* of v from u, and it obeys $h_{i,i} = \frac{1}{\Pi_i}$ for an ergodic chain with stationary distribution Π.

4.4.1   Random Walks on Undirected Graphs

We consider a random walk X on a graph G as before, but now with the premise that G is undirected.

Clearly, X will be irreducible iff G is connected. It can also be shown that it will be aperiodic iff G is not bipartite. The $\Rightarrow$ direction follows from the fact that paths between two sides of a bipartite graph are always of even length, whereas the $\Leftarrow$ direction follows from the fact that a non-bipartite graph always contains a cycle of odd length.

We can always make a walk on a connected graph ergodic simply by adding self-loops to one or more of the vertices.

4.4.1.1   Ergodic Random Walks on Undirected Graphs

Theorem 4.3 *If the random walk X on G is ergodic, then its stationary distribution Π is given by* $\forall v \in V, \Pi_v = \frac{d(v)}{2m}$.

Proof Let Π be as defined above. Then:

$$(\Pi P)_v = \sum_{u,v \in E} \Pi_u \frac{1}{d(u)}$$

$$= \sum_{u,u,v \in E} \frac{1}{2m}$$

$$= \frac{d(v)}{2m}$$

$$= \Pi_v$$

So as $\sum_v \Pi_v = \frac{2m}{2m} = 1$, Π is the stationary distribution of X.

In general, even on this subset of random walks, the hitting time will not be symmetric, as will be shown in our next example. So we define the commute time $C_{u,v} = h_{u,v} + h_{v,u}$.

4.4.2 Electric Networks and Random Walks

A resistive electrical network is an undirected graph; each edge has branch resistance associated with it. The electrical flow is determined by two laws: Kirchhoff's law (preservation of flow - all the flow coming into a vertex, leaves it) and Ohm's law (the voltage across a resistor equals the product of the resistance times the current through it). View graph G as an electrical network with unit resistors as edges. Let $R_{u,v}$ be the effective resistance between vertices u and v. The commute time between u and v in a graph is related to $R_{u,v}$ by $C_{u,v} = 2mR_{u,v}$. We get the following inequalities assuming this relation.

If $(u, v) \in E$,

$$R_{u,v} \le 1 \therefore C_{u,v} \le 2m$$

In general, $\forall u, v \in V$,

$$R_{u,v} \le n - 1 \therefore C_{u,v} \le 2m(n - 1) < n^3$$

We inject $d(v)$ amperes of current into $\forall v \in V$. Eventually, we select some vertex $u \in V$ and remove $2m$ current from u leaving net $d(u) - 2m$ current at u. Now we get voltages x_v $\forall v \in V$. Suppose we have $x_v - x_u = h_{v,u}$ $\forall v \ne u \in V$. Let L be the Laplacian for G and D be the degree vector, then we have

$$Lx = i_u = D - 2m\mathbb{1}_u$$

$$\forall v \in V, \sum_{(u,v) \in E} x_v - x_u = d(v) \tag{4.6}$$

You might now see the connection between a random walk on a graph and electrical network. Intuitively, the electricity, is made out of electrons each one of them is doing a random walk on the electric network. The resistance of an edge, corresponds to the probability of taking the edge.

4.4.3 Example: The Lollipop Graph

This is one example of a graph where the cover time depends on the starting vertex. The lollipop graph on n vertices is a clique of $\frac{n}{2}$ vertices connected to a path of $\frac{n}{2}$

vertices. Let u be any vertex in the clique that does not neighbour a vertex in the path, and v be the vertex at the end of the path that does not neighbour the clique. Then $h_{u,v} = \theta(n^3)$ while $h_{v,u} = \theta(n^2)$. This is because it takes $\theta(n)$ time to go from one vertex in the clique to another, and $\theta(n^2)$ time to successfully proceed up the path, but when travelling from u to v the walk will fall back into the clique $\theta(1)$ times as often as it makes it a step along the path to the right, adding an extra factor of n to the hitting time.

To compute $h_{u,v}$. Let u' be the vertex common to the clique and the path. Clearly, the path has resistance $\theta(n)$. $\theta(n)$ current is injected in the path and $\theta(n^2)$ current is injected in the clique.

Consider draining current from v. The current in the path is $\theta(n^2)$ as $2m - 1 = \theta(n^2)$ current is drained from v which enters v through the path implying $x'_u - x_v = \theta(n^3)$ using Ohm's law ($V = IR$). Now consider draining current from u instead. The current in the path is now $\theta(n)$ implying $x_v - x'_u = \theta(n^2)$ by the same argument.

Since the effective resistance between any edge in the clique is less than 1 and $\theta(n^2)$ current is injected, there can be only $\theta(n^2)$ voltage gap between any 2 vertices in the clique. We get $h_{u,v} = x_u - x_v = \theta(n^3)$ in the former case and $h_{v,u} = x_v - x_u = \theta(n^2)$ in the latter.

4.5   Crowdsourcing Model

Crowdsourcing techniques are very powerful when harnessed for the purpose of collecting and managing data. In order to provide sound scientific foundations for crowdsourcing and support the development of efficient crowdsourcing processes, adequate formal models must be defined. In particular, the models must formalize unique characteristics of crowd-based settings, such as the knowledge of the crowd and crowd-provided data; the interaction with crowd members; the inherent inaccuracies and disagreements in crowd answers; and evaluation metrics that capture the cost and effort of the crowd.

To work with the crowd, one has to overcome several challenges, such as dealing with users of different expertise and reliability, and whose time, memory and attention are limited; handling data that is uncertain, subjective and contradictory; and so on. Particular crowd platforms typically tackle these challenges in an ad hoc manner, which is application-specific and rarely sharable. These challenges along with the evident potential of crowdsourcing have raised the attention of the scientific community, and called for developing sound foundations and provably efficient approaches to crowdsourcing. In cases where the crowd is utilised to filter, group or sort the data, standard data models can be used. The novelty here lies in cases when some of the data is harvested with the help of the crowd. One can generally distinguish between procuring two types of data: general data that captures truth that normally resides in a standard database, for instance, the locations of places or opening hours; versus individual data that concerns individual people, such as their preferences or habits.

4.5.1 Formal Model

We now present a combined formal model for the crowd mining setting of (Amarilli et al. 2014; Amsterdamer et al. 2013).

Let $I = \{i_1, i_2, i_3, \ldots\}$ be a finite set of item names. Define a *database* D as a finite bag (multiset) of *transactions* over I, s.t. each transaction $T \in D$ represents an occasion, e.g., a meal. We start with a simple model where every T contains an itemset $A \subseteq I$, reflecting, e.g., the set of food dishes consumed in a particular meal. Let U be a set of users. Every $u \in U$ is associated with a *personal database* D_u containing the transactions of u (e.g., all the meals in u's history). $|D_u|$ denotes the number of transactions in D_u. The frequency or *support* of an itemset $A \subseteq I$ in D_u is $\mathrm{supp}_u(A) := |\{T \in D_u | A \subseteq T\}|/|D_u|$. This individual significance measure will be aggregated to identify the overall frequent itemsets in the population. For example, in the domain of culinary habits, I may consist of different food items. A transaction $T \in D_u$ will contain all the items in I consumed by u in a particular meal. If, for instance, the set {tea, biscuits, juice} is frequent, it means that these food and drink items form a frequently consumed combination.

There can be dependencies between *itemsets* resulting from semantic relations between *items*. For instance, the itemset {cake, tea} is semantically implied by any transaction containing {cake, jasmine tea}, since jasmine tea is a (kind of) tea. Such semantic dependencies can be naturally captured by a *taxonomy*. Formally, we define a taxonomy Ψ as a partial order over I, such that $i \leq i'$ indicates that item i' is more specific than i (any i' is also an i).

Based on $\leq$, the semantic relationship between items, we can define a corresponding order relation on itemsets.[1] For itemsets A, B we define $A \leq B$ iff every item in A is implied by some item in B. We call the obtained structure the *itemset taxonomy* and denote it by $I(\Psi)$. $I(\Psi)$ is then used to extend the definition of the support of an itemset A to $\mathrm{supp}_u(A) := |\{T \in D_u | A \leq T\}|/|D_u|$, i.e., the fraction of transactions that *semantically imply* A.

Reference (Amarilli et al. 2014) discusses the feasibility of *crowd-efficient* algorithms by using the computational complexity of algorithms that achieve the upper crowd complexity bound. In all problem variants, they have the crowd complexity lower bound as a simple lower bound. For some variants, they illustrated that, even when the crowd complexity is feasible, the underlying computational complexity may still be infeasible.

---

[1] Some itemsets that are semantically equivalent are identified by this relation, e.g., {tea, jasmine tea} is represented by the equivalent, more concise {jasmine tea} because drinking jasmine tea is a simply case of drinking tea.

4.6 Communication Complexity

Communication complexity explores how much two parties need to communicate in order to compute a function whose output depends on information distributed over both parties. This mathematical model allows communication complexity to be applied in many different situations, and it has become an key component in the theoretical computer science toolbox. In the communication setting, Alice has some input x and Bob has some input y. They share some public randomness and want to compute $f(x, y)$. Alice sends some message m_1, and then Bob responds with m_2, and then Alice responds with m_3, and so on. At the end, Bob outputs $f(x, y)$. They can choose a protocol Π, which decides how to assign what you send next based on the messages you have seen so far and your input. The total number of bits transfered is $|\Pi| = \sum |m_i|$.

The communication complexity of the protocol Π is

$$CC_\mu(\Pi) = e_\mu(|\Pi|),$$

where μ is a distribution over the inputs (x, y) and the protocol. The communication complexity of the function f for a distribution μ is

$$CC_\mu(f) = \min_{\Pi \text{ solves } f \text{ with } 3/4 \text{ prob}} CC_\mu(\Pi).$$

The communication complexity of the function f is

$$CC(f) = \max_\mu CC_\mu(f).$$

4.6.1 Information Cost

Information cost is related to communication complexity, as entropy is related to compression.

Recall that the entropy is $H(X) = \sum p(x) \log \frac{1}{p(x)}$. Now, the mutual information $I(X; Y) = H(X) - H(X|Y)$ between X and Y is how much a variable Y tells you about X. It is actually interesting that we also have $I(X; Y) = H(Y) - H(Y|X)$.

The information cost of a protocol Π is

$$IC(\Pi) = I(X; \Pi|Y) + I(Y; \Pi|X).$$

This is how much Bob learns from the protocol about X plus how much Alice learns from the protocol about Y. The information cost of a function f is

$$IC(f) = \min_{\Pi \text{ solves } f} IC(\Pi).$$

For all protocol Π, we have $IC(\Pi) \le e|\Pi| = CC(\Pi)$, because there are at most b bits of information if there are only b bits transmitted in the protocol. Taking the minimum over all protocols implies $IC(f) \le CC(f)$. This is analogous to Shannon's result that $H \le \ell$.

It is really interesting that the asymptotic statement is true. Suppose we want to solve n copies of the communication problem. Alice given $x_1, \ldots, x_n$ and Bob given $y_1, \ldots, y_n$, they want to solve $f(x_1, y_1), \ldots, f(x_n, y_n)$, each failing at most $1/4$ of the time. We call this problem the direct sum $f^{\oplus n}$. Then, for all functions f, it is not hard to show that $IC(f^{\oplus n}) = nIC(f)$.

Theorem 4.4 (Braverman and Rao 2011)

$$\frac{CC(f^{\oplus n})}{n} \to IC(f) \quad as\ n \to \infty.$$

In the limit, this theorem suggests that information cost is the right notion.

4.6.2 Separation of Information and Communication

The remaining question is, for a single function, whether $CC(f) \approx IC(f)$, in particular whether $CC(f) = IC(f)O(1) + O(1)$. If this is true, it would prove the direct sum conjecture $CC(f^{\oplus n}) \gtrsim nCC(f) - O(1)$.

The recent paper by Ganor, Kol and Raz (Ganor et al. 2014) showed that it is not true. They gave a function f for which $IC(f) = k$ and $CC(f) \ge 2^{\Omega(k)}$. This is the best because it was known before this that $CC(f) \le 2^{O(IC(f))}$. The function that they gave has input size $2^{2^{2^k}}$. So, it is still open whether $CC(f) \lesssim IC(f) \log \log |\text{input}|$.

A binary tree with depth 2^{2^k} is split into levels of width $\approx k$. For every vertex v in the tree, there are two associated values x_v and y_v. There is a random special level of width $\approx k$. Outside this special level, we have $x_v = y_v$ for all v. We think about x_v and y_v as which direction you ought to go. So, if they are both 0, you want to go in one direction. If they are both 1, you want to go in the other. Within the special level, the values x_v and y_v are uniform. At the bottom of the special level, v is *good* if the path to v is following directions. The goal is to agree on any leaf v' where v' is a descendent of some good vertex.

Here we do not know where the special level is, because if you knew where the special level was, then $O(k)$ communication suffices. The problem is you do not know where the special level is. You can try binary searching to find the special level, taking $O(2^k)$ communication. This is basically the best you can do apparently.

We can construct a protocol with information cost only $O(k)$. It is okay to transmit something very large as long as the amount of information contained in it is small. Alice can transmit her path and Bob just follows it, and that is a large amount of communication but it is not so much information because Bob knows what the first set would be. The issue is that it still gives you $\approx 2^k$ bits of information knowing where the special level is. The idea is instead that Alice chooses a noisy path where

90% of the time follows her directions and 10% deviates. This path is transmitted to Bob. It can be shown that this protocol only has $O(k)$ information. Therefore, many copies can get more efficient.

4.7   Adaptive Sparse Recovery

Adaptive sparse recovery is like the conversation version of sparse recovery.

In non-adaptive sparse recovery, Alice has $i \in [n]$ and sets $x = e_i + w$. She transmits $y = Ax = Ae_i + w'$. Bob receives y and recovers $y \to \hat{x} \approx x \to \hat{i} \approx i$. In this one-way conversation,

$$
\begin{aligned}
I(\hat{i}; i) &\leq I(y; i) \\
&\leq m(0.5 \log(1 + \text{SNR})) \\
&\lesssim m \\
H(i) - H(i|\hat{i}) &\lesssim m \\
\log n - (0.25 \log n + 1) &\lesssim m \\
m &\gtrsim \log n.
\end{aligned}
$$

In the adaptive case, we have something more of a conversation. Alice knows x. Bob sends v_1 and Alice sends back $\langle v_1, x \rangle$. Then, Bob sends v_2 and Alice sends back $\langle v_2, x \rangle$. And then, Bob sends v_3 and Alice sends back $\langle v_3, x \rangle$, and so on.

To show a lower bound, consider stage r. Define P as the distribution of $(i|y_1, \ldots, y_{r-1})$. Then, the observed information by round r is $b = \log n - H(P) = e_{i \sim P} \log(np_i)$. For a fixed v depending on P, as $i \sim P$, we know that

$$
I(\langle v, x \rangle; i) \leq \frac{1}{2} \log \left(1 + \frac{e_{i \sim P} v_i^2}{\|v_i\|_2^2 / n} \right).
$$

With some algebra (Lemma 3.1 in (Price and Woodruff 2013)), we can bound the above expression by $O(b + 1)$. It means that on average the number of bits that you get at the next stage is $\lesssim 2$ times what you had at the previous stage. This implies that R rounds take $\Omega(R \log^{1/R} n)$ measurements. And in general, it takes $\Omega(\log \log n)$ measurements.

References

Achlioptas D (2003) Database-friendly random projections. J Comput Syst Sci 66(4):671–687

Ahn KJ, Guha S, McGregor A (2012) Analyzing graph structure via linear measurements. SODA 2012:459–467

Ailon N, Chazelle B (2009) The fast Johnson-Lindenstrauss transform and approximate nearest neighbors. SIAM J Comput 39(1):302–322

Alon N (2003) Problems and results in extremal combinatorics-I. Discret Math 273(1–3):31–53

Alon N, Matias Y, Szegedy M (1999) The space complexity of approximating the frequency moments. J Comput Syst Sci 58(1):137–147

Amarilli A, Amsterdamer Y, Milo T (2014) On the complexity of mining itemsets from the crowd using taxonomies. ICDT

Amsterdamer Y, Grossman Y, Milo T, Senellart P (2013) Crowd mining. SIGMOD

Andoni A (2012) High frequency moments via max-stability. Manuscript

Andoni A, Krauthgamer R, Onak K (2011) Streaming algorithms via precision sampling. FOCS:363–372

Avron H, Maymounkov P, Toledo S (2010) Blendenpik: Supercharging LAPACK's least-squares solver. SIAM J Sci Comput 32(3):1217–1236

Bahmani B, Kumar R, Vassilvitskii S (2012) Densest subgraph in streaming and mapreduce proc. VLDB Endow 5(5):454–465

Bar-Yossef Z, Jayram TS, Kumar R, Sivakumar D (2004) An information statistics approach to data stream and communication complexity. J Comput Syst Sci 68(4):702–732

Beame P, Fich FE (2002) Optimal bounds for the predecessor problem and related problems. JCSS 65(1):38–72

Braverman M, Rao A (2011) Information equals amortized communication. FOCS 2011:748–757

Brinkman B, Charikar M (2005) On the impossibility of dimension reduction in l_1. J ACM 52(5):766–788

Candès EJ, Tao T (2010) The power of convex relaxation: near-optimal matrix completion. IEEE Trans Inf Theory 56(5):2053–2080

Candès EJ, Romberg JK, Tao T (2006) Robust uncertainty principles: exact signal reconstruction from highly incomplete frequency information. IEEE Trans Inf Theory 52(2):489–509

Chakrabarti A, Khot S, Sun X (2003) Near-optimal lower bounds on the multi-party communication complexity of set disjointness. In: IEEE conference on computational complexity, pp 107–117

Chakrabarti A, Shi Y, Wirth A, Chi-Chih Yao A (2001) Informational complexity and the direct sum problem for simultaneous message complexity. FOCS:270–278

Charikar M, Chen K, Farach-Colton M (2002) Finding frequent items in data streams. ICALP 55(1)

Clarkson KL, Woodruff DP (2013) Low rank approximation and regression in input sparsity time. In: Proceedings of the 45th annual ACM symposium on the theory of computing (STOC), pp 81–90

Cormen TH, Leiserson CE, Rivest RL, Stein C (2009). Introduction to algorithms. MIT Press

Cormode G, Muthukrishnan S (2005) An improved data stream summary: the count-min sketch and its applications. J Algorithms 55(1):58–75

Dasgupta A, Kumar R, Sarlós T (2010) A sparse Johnson: Lindenstrauss transform. STOC:341–350

Dean J, Ghemawat S (2004) MapReduce: Simplified data processing on large clusters. In: proceedings of the sixth symposium on operating system design and implementation. (San Francisco, CA, Dec 6–8). Usenix Association

Demmel J, Dumitriu I, Holtz O (2007) Fast linear algebra is stable. Numer Math 108(1):59–91

Dirksen S (2015) Tail bounds via generic chaining. Electron J Probab 20(53):1–29

Donoho DL (2006) Compressed sensing. IEEE Trans Inf Theory 52(4):1289–1306

Drineas P, Mahoney MW, Muthukrishnan S (2006) Sampling algorithms for l_2 regression and applications. SODA 2006:1127–1136

Emmanuel J (2009) Candès and Benjamin Recht. Exact matrix completion via convex optimization. Found Comput Math 9(6):717–772

Feigenbaum J, Kannan S, McGregor A, Suri S, Zhang J (2005) On graph problems in a semi-streaming model. Theor Comput Sci 348(2–3):207–216

Fernique X (1975) Regularité des trajectoires des fonctions aléatoires gaussiennes. Ecole d'Eté de Probabilités de Saint-Flour IV, Lecture Notes in Math 480:1–96

Fredman ML, Komlós J, Szemerédi E (1984) Storing a sparse table with O(1) worst case access time. JACM 31(3):538–544

Frieze AM, Kannan R, Vempala S (2004) Fast Monte-Carlo algorithms for finding low-rank approximations. J ACM 51(6):1025–1041

Ganor A, Kol G, Raz R (2014) Exponential separation of information and communication. ECCC, Revision 1 of Report No. 49

Globerson A, Chechik G, Tishby N (2003) Sufficient dimensionality reduction with irrelevance statistics. In: Proceeding of the 19th conference on uncertainty in artificial intelligence, Acapulco, Mexico

Gordon Y ((1986–1987)) On Milman's inequality and random subspaces which escape through a mesh in R^n. In: Geometric aspects of functional analysis vol 1317:84–106

Gronemeier A (2009) Asymptotically optimal lower bounds on the NIH-multi-party information complexity of the AND-function and disjointness. STACS, pp 505–516

Gross D (2011) Recovering low-rank matrices from few coefficients in any basis. IEEE Trans Inf Theory 57:1548–1566

Gross D, Liu Y-K, Flammia ST, Becker S, Eisert J (2010) Quantum state tomography via compressed sensing. Phys Rev Lett 105(15):150401

Guha S, McGregor A (2012) Graph synopses, sketches, and streams: a survey. PVLDB 5(12):2030–2031

Guyon I, Gunn S, Ben-Hur A, Dror G (2005) Result analysis of the NIPS 2003 feature selection challenge. In: Neural information processing systems. Curran & Associates Inc., Red Hook

Hanson DL, Wright FT (1971) A bound on tail probabilities for quadratic forms in independent random variables. Ann Math Stat 42(3):1079–1083

Hardt M (2014) Understanding alternating minimization for matrix completion. FOCS:651–660

Hardt M, Wootters M (2014) Fast matrix completion without the condition number. COLT:638–678

Indyk P (2003) Better algorithms for high-dimensional proximity problems via asymmetric embeddings. In: ACM-SIAM symposium on discrete algorithms

Indyk P (2006) Stable distributions, pseudorandom generators, embeddings, and data stream computation. J. ACM 53(3):307–323

Indyk P, Woodruff DP (2005) Optimal approximations of the frequency moments of data streams. STOC:202–208

Jayram TS (2009) Hellinger strikes back: a note on the multi-party information complexity of AND. APPROX-RANDOM, pp 562–573

Jayram TS, Woodruff DP (2013) Optimal bounds for Johnson-Lindenstrauss transforms and streaming problems with subconstant error. ACM Trans Algorithms 9(3):26

Johnson WB, Lindenstrauss J (1984) Extensions of Lipschitz mappings into a Hilbert space. Contemp Math 26:189–206

Johnson WB, Naor A (2010) The Johnson-Lindenstrauss lemma almost characterizes Hilbert space, but not quite. Discret Comput Geom 43(3):542–553

Jowhari H, Saglam M, Tardos G (2011) Tight bounds for L_p samplers, finding duplicates in streams, and related problems. PODS 2011:49–58

Kane DM, Meka R, Nelson J (2011) Almost optimal explicit Johnson-Lindenstrauss transformations. In: Proceedings of the 15th international workshop on randomization and computation (RANDOM), pp 628–639

Kane DM, Nelson J (2014) Sparser Johnson-Lindenstrauss transforms. J ACM 61(1):4:1–4:23

Kane DM, Nelson J, Woodruff DP (2010) An optimal algorithm for the distinct elements problem. In: Proceedings of the twenty-ninth ACMSIGMOD-SIGACT-SIGART symposium on principles of database systems (PODS), pp 41–52

Karp RM, Vazirani UV, Vazirani VV (1990) An optimal algorithm for on-line bipartite matching. In: STOC '90: Proceedings of the twenty-second annual ACM symposium on theory of computing. ACM Press, New York, pp 352–358

Keshavan RH, Montanari A, Oh S (2010) Matrix completion from noisy entries. J Mach Learn Res 99:2057–2078

Klartag B, Mendelson S (2005) Empirical processes and random projections. J Funct Anal 225(1):229–245

Kushilevitz E, Nisan N (1997) Communication complexity. Cambridge University Press, Cambridge

Larsen KG, Nelson J (2014) The Johnson-Lindenstrauss lemma is optimal for linear dimensionality reduction. In: CoRR. arXiv:1411.2404

Larsen KG, Nelson J, Nguyen HL (2014) Time lower bounds for nonadaptive turnstile streaming algorithms. In: CoRR. arXiv:1407.2151

Lévy P (1925) Calcul des probabilités. Gauthier-Villars, Paris

Lloy S (1982) Least squares quantization in PCM. IEEE Trans Inf Theory 28(2):129–137

Lust-Piquard F, Pisier G (1991) Non commutative Khintchine and Paley inequalities. Arkiv för Matematik 29(1):241–260

Matousek J (2008) On variants of the Johnson-Lindenstrauss lemma. Random Struct Algorithms 33(2):142–156

Mendelson S, Pajor A, Tomczak-Jaegermann N (2007) Reconstruction and subgaussian operators in asymptotic geometric analysis. Geom Funct Anal 1:1248–1282

Motwani R, Raghavan P (1995) Randomized algorithms. Cambridge University Press, Cambridge, pp 0–521-47465-5

Nelson J (2015) CS 229r: Algorithms for big data. Course, Web, Harvard

Nelson J, Nguyen HL, Woodruff DP (2014) On deterministic sketching and streaming for sparse recovery and norm estimation. Linear algebra and its applications, special issue on sparse approximate solution of linear systems. 441:152–167

Nisan N (1992) Pseudorandom generators for space-bounded computation. Combinatorica 12(4):449–461

Oymak S, Recht B, Soltanolkotabi M (2015) Isometric sketching of any set via the restricted isometry property. In: CoRR. arXiv:1506.03521

Papadimitriou CH, Raghavan P, Tamaki H, Vempala S (2000) Latent semantic indexing: a probabilistic analysis. J Comput Syst Sci 61(2):217–235

Price E, Woodruff DP (2013) Lower bounds for adaptive sparse recovery. SODA 2013:652–663

Recht B (2011) A simpler approach to matrix completion. J Mach Learn Res 12:3413–3430

Recht B, Fazel M, Parrilo PA (2010) Guaranteed minimum-rank solutions of linear matrix equations via nuclear norm minimization. SIAM Rev 52(3):471–501

Rokhlin V, Tygert M (2008) A fast randomized algorithm for overdetermined linear least-squares regression. Proc Natl Acad Sci 105(36):13212–13217

Rubinfeld R (2009) Sublinear time algorithms. Tel-Aviv University, Course, Web

Sarlós T (2006) Improved approximation algorithms for large matrices via random projections. In: 47th annual IEEE symposium on foundations of computer science FOCS:143–152

Sarlós T, Benczúr AA, Csalogány K, Fogaras D, Rácz B (2006) To randomize or not to randomize: space optimal summarise for hyperlink analysis. In: International conference on world wide web (WWW)

Schramm T, Weitz B (2015) Low-rank matrix completion with adversarial missing entries. In: CoRR. arXiv:1506.03137

Talagrand M (1996) Majorizing measures: the generic chaining. Ann Probab 24(3):1049–1103

Wright SJ, Nowak RD, Figueiredo MAT (2009) Sparse reconstruction by separable approximation. IEEE Trans Signal Process 57(7):2479–2493

Printed in the United States
By Bookmasters